FRACKING UNCERTAINTY

Hydraulic Fracturing and the Provincial Politics of Risk

Hydraulic fracturing – *fracking* – is an unconventional extraction technique used in the oil and gas industry that has fundamentally transformed global energy politics. In *Fracking Uncertainty*, Heather Millar explains variation in Canadian provincial policy approaches, which range from pro-development regulation to moratoria and outright bans. Millar argues that although regulatory designs are shaped by governments' desires to seek out economic benefits or protect against environmental harms, policy makers' perceptions of said benefits and/ or harms are mediated through socially constructed narratives about uncertainty and risk.

Fracking Uncertainty offers in-depth case studies of regulatory development in British Columbia, Alberta, New Brunswick, and Nova Scotia. Drawing on media analysis and interviews with government officials, industry representatives, academics, and environmental advocates, Millar demonstrates how risk narratives foster distinctive forms of learning in each province, leading to different regulatory reforms.

(Studies in Comparative Political Economy and Public Policy)

HEATHER MILLAR is an associate professor in the Department of Political Science at the University of New Brunswick.

Studies in Comparative Political Economy and Public Policy

Editors: Michael Howlett, David Laycock (Simon Fraser University), and Stephen Mcbride (McMaster University)

Studies in Comparative Political Economy and Public Policy is designed to showcase innovative approaches to political economy and public policy from a comparative perspective. While originating in Canada, the series will provide attractive offerings to a wide international audience, featuring studies with local, subnational, cross-national, and international empirical bases and theoretical frameworks.

For a list of books published in the series, see page 247.

Fracking Uncertainty

Hydraulic Fracturing and the Provincial Politics of Risk

HEATHER MILLAR

UNIVERSITY OF TORONTO PRESS
Toronto Buffalo London

ISBN 978-1-4875-5268-8 (cloth)   ISBN 978-1-4875-5270-1 (EPUB)
ISBN 978-1-4875-5269-5 (paper)   ISBN 978-1-4875-5271-8 (PDF)

Library and Archives Canada Cataloguing in Publication
Title: Fracking uncertainty : hydraulic fracturing and the provincial politics of risk /
    Heather Millar.
Names: Millar, Heather, author.
Series: Studies in comparative political economy and public policy.
Description: Series statement: Studies in comparative political economy and
    public policy | Includes bibliographical references and index.
Identifiers: Canadiana (print) 20240388933 | Canadiana (ebook) 20240388968 |
    ISBN 9781487552688 (cloth) | ISBN 9781487552695 (paper) | ISBN 9781487552701
    (EPUB) | ISBN 9781487552718 (PDF)
Subjects: LCSH: Hydraulic fracturing – Government policy – British Columbia –
    Case studies. | LCSH: Hydraulic fracturing – Government policy – Alberta –
    Case studies. | LCSH: Hydraulic fracturing – Government policy –
    New Brunswick – Case studies. | LCSH: Hydraulic fracturing –
    Government policy – Nova Scotia – Case studies.
Classification: LCC TN871.255 .M55 2024 | DDC 622/.3380971 – dc23

Cover design: Heng Wee Tan
Cover image: tbob / istockphoto.com

We wish to acknowledge the land on which the University of Toronto Press
operates. This land is the traditional territory of the Wendat, the Anishnaabeg,
the Haudenosaunee, the Métis, and the Mississaugas of the Credit First Nation.

This book has been published with the help of a grant from the Federation
for the Humanities and Social Sciences, through the Awards to Scholarly
Publications Program, using funds provided by the Social Sciences and
Humanities Research Council of Canada.

University of Toronto Press acknowledges the financial support of the
Government of Canada, the Canada Council for the Arts, and the Ontario Arts
Council, an agency of the Government of Ontario, for its publishing activities.

Canada Council   Conseil des Arts
for the Arts   du Canada

ONTARIO ARTS COUNCIL
CONSEIL DES ARTS DE L'ONTARIO
an Ontario government agency
un organisme du gouvernement de l'Ontario

Funded by the   Financé par le
Government   gouvernement
of Canada   du Canada

Canadä

Contents

Tables and Figures

Tables

Figures

Acknowledgments

My journey into the intricacies of energy politics research has been rife with its own uncertainties, complexities, and unknown unknowns; I am very grateful to have been graced with many supportive companions along the way.

The beginnings of *Fracking Uncertainty* were shaped by my doctoral studies at the Department of Political Science at University of Toronto, where I was fortunate to find a welcoming home in the sub-field of policy studies. My doctoral supervisor, Linda A. White, has been an enthusiastic and dedicated champion; her intellectual curiosity and collaborative spirit are ongoing sources of inspiration and growth as I wade further and further into academic life. I am very grateful for the expansive knowledge, insightful guidance, and kind humour of committee members Grace Skogstad, Andrea Olive, Kate Neville, and external examiner Kathryn Harrison, each of whom dedicated countless hours helping me clarify my thoughts and refine my arguments. Appreciation is also due to the vibrant community of scholars at the Environmental Governance Lab, where I first presented initial findings and later completed a postdoctoral fellowship. Special thanks to my postdoctoral supervisors Matthew Hoffmann and Steven Bernstein for encouraging me to think of myself as an environmental politics scholar, and to Teresa Kramarz, Sara Hughes, Laura Tozer, and Stefan Renckens for providing perceptive comments at various stages of the project. I extend my warmest thanks to all the interview participants for their contributions to my study; thank you for your time and insight in teasing out the complexities of hydraulic fracturing in Canada. I would also like to acknowledge the generous funding received from the Canada Graduate Scholarship Program of the Social Science and Humanities Research Council, the Ontario Graduate Scholarship, the Department of Political Science, and the School of Graduate Studies at the University of Toronto that made my doctoral studies and my field research possible as well as to the Harrison McCain Foundation for publication support in the final year of the project.

Working on hydraulic fracturing regulation occasionally felt like a winding road around the margins of political science; I am extremely grateful to comparative public policy scholars Barry Rabe, Christopher Gore, Erick Lachapelle, Éric Montpetit, Brendan Boyd, Angela Carter, and George Hoberg for kindly welcoming me into the world of climate and energy politics and for their generous feedback on various conference papers and chapter drafts. Many thanks to Monica Gattinger and the Institute for Science, Society, and Policy at the University for Ottawa for hosting me as a SSHRC Postdoctoral Fellow and for deftly demonstrating how to translate research into practical insights for policy makers and regulators.

Both the book (and I!) matured in the supportive environment of the Department of Political Science at the University of New Brunswick. My colleagues Catherine Bigonnesse, Herb Emery, Zabrina Hamilton, Paul Howe, George MacLean, Ted McDonald, Thom Workman, and Don and Joanne Wright have been invaluable as I have adjusted to the new and uncertain challenges of faculty life. Special thanks to Suzanne Hindmarch, Biljana Stevanovski, Antulio Rosales, and Tobin Haley for providing kind commiserations, walks in the woods, and encouragement in times of need. As a new faculty member, I have benefited immensely from the mentorship of senior colleagues at other institutions: Alana Cattapan, Ethel Tungohan, Beth Schwartz, Sarah Martin, Isabelle Côté, and Alison Smith have all provided me with gems of inspiration and a wealth of pedagogical resources along the way. Many thanks to Chelsea Pennell (at the University of Toronto), and Samantha McRae (at UNB) for their efficient research assistance coding media articles for the project.

I am extremely grateful to Daniel Quinlan, my editor at the University of Toronto Press, for his early championship of my work and extreme patience in shepherding the book to completion. I wish that every new writer could have such a supportive and encouraging guide into the world of academic book publishing. My sincere appreciations to the peer reviewers for their time, labour, and constructive engagement with the book: any (uncertain) omissions or (catastrophic) mistakes in the narrative are mine alone.

One of the enduring joys of becoming an academic has been the discovery of intellectual kindred spirits and delightful friends. Fellow graduate students Carmen Ho, Marion Laurence, Matthew Lesch, Hamish van der Ven, Carey Doberstein, Adrienne Davidson, Andrew

McDougall, Eve Borgeois, Minh Do, and Sophie Borwein have provided me endless feedback and support, generously listening to long rambles, reading drafts, and formatting power point presentations, long past graduation. To Amy Janzwood, co-author, collaborator, and friend, thank you for our many discussions about the interpretive politics of natural gas, the intricacies of carbon lock-in, and the urgent need for justice. I look forward to our continued conversations, both online and in person.

My friends and family have been crucial to ensuring my perseverance and maintaining my mental health over the many years of working on this project. Many thanks to dear friends Erin Case, Kyla Epstein, and Caitlyn Vernon for encouraging me through years of imposter syndrome and to Katy Konyk, Melanie Goedecki, and Hannah Classen for providing much needed friendship outside of the confines of academia. Many thanks to my siblings and in-laws for their support, with a special shout-out to my sister-in-law Shannon Whibbs, for her unflagging enthusiasm and professional editorial smarts.

The dream of writing an academic book would have never become a reality without the unfailing support of my parents Robin and John Millar. Their deep commitment to lifelong learning was evident throughout my childhood and their complete support of my choice to leave a successful career to pursue graduate studies was never in doubt. My mother Robin demonstrated to me the possibility of returning to school as a mature student by obtaining a doctorate in her fifties. I aspire to her thoughtful leadership and delight in intellectual exploration. My father John was my biggest coach and cheerleader. Although he passed away in the final years of my doctoral studies, I know he would be absolutely chuffed to see *Fracking Uncertainty* in print, if only to appreciate the pun.

My research into the uncertainties of hydraulic fracturing would have been infinitely less enjoyable without the companionship of my partner Christopher Whibbs. Chris had the temerity to start dating during my comprehensive exams, to stick with it through the dark days of proposal writing, to marry in the middle of my field research and then upend our entire life and move to Fredericton – amid a global pandemic no less – to support my dream of becoming a professor. His unwavering support as I have navigated the daily highs and lows of research, writing, and teaching has been without compare. His infallible commitment to my success has enabled me to "get'r done" especially after the arrival of our daughter Bridget. Although her spunky presence in our lives has likely prolonged this book's time to completion, every day with her has been a certain source of learning and joy.

FRACKING UNCERTAINTY

Chapter One

Fracking and the Politics of Risk

In September 2014 reporters on the campaign trail for the New Brunswick provincial election asked the incumbent Progressive Conservative premier, David Alward, to comment on the ban on high-volume hydraulic fracturing[1] that had been recently announced in the neighbouring province of Nova Scotia. The request to comment on another jurisdiction's governance was not as strange as it might seem. Since 2011 shale gas development had become a thorny topic for the Alward government and the management of hydraulic fracturing was gaining ground as a key election issue. Premier Alward responded definitively, noting "if Nova Scotia is saying no to the development of our natural resources, including shale gas, then they're saying no to becoming a have economy and a have province. That's a decision that the government there has to make" (Berry 2014a). For Alward, the ban represented a flagrant failure to develop a healthy economy, a core responsibility of government.

In contrast, New Brunswick Liberal Party challenger Brian Gallant advocated for a precautionary approach, arguing that "provinces feel hydraulic fracturing should not be done right now because there just isn't enough information about the risks, how to mitigate those risks, how to regulate those risks and how to enforce those regulations" (Berry 2014a). For Gallant, the ban was a mechanism to reconcile the lack of knowledge regarding uncertain environmental harms and to move toward responsible government.

Back in Nova Scotia, Energy Minister Andrew Younger put forward a more populist rationale for the government's action, stating "there is not a community in this province ... where there's a large number of people pushing to allow hydraulic fracturing ... the resources belong to the people of Nova Scotia, and they get to decide how they are harnessed" (Berry 2014a). For Younger, the ban was a method by which to entrench the will of the people into law.

Implicitly, these different rationales supported alternate policy solutions: for Alward development was a foregone conclusion, for Gallant a moratorium allowed for future research into a new technology, and for Younger the ban reflected the lack of acceptance of the technology among the electorate. How is it that these provincial decision makers came to such different accounts of the same technology?

The regional debate described above is one of a multitude of policy debates that erupted across North America in the early 2010s regarding the technological practice of hydraulic fracturing, an unconventional extraction technique used in the oil and gas industry. Although hydraulic fracturing has been used in the industry for decades to access conventional resources on a small scale (Natural Resources Canada 2015), since the early 2000s producers have successfully adapted the technology to efficiently extract oil and gas from previously untenable resources, including deposits of shale rock. This method of industrial production, commonly known as "fracking," has fundamentally transformed global energy markets, shifting the United States from being an importer to being an exporter of natural gas extracted from shale, and having corresponding impacts on Canadian energy markets (US EIA 2014; 2017; 2019; Gattinger and Aguirre 2016). The boom in US natural gas production through hydraulic fracturing had a significant impact on domestic and global market dynamics, generating volatility in domestic prices of natural gas as well as affecting European Union coal and global markets for liquefied natural gas (Foss 2011; Ernst & Young 2013; Neville et al. 2017). In both the United States and Europe, governments and industry proponents stressed the potential for natural gas to shift global markets to a lower-carbon economy, by facilitating a shift away from more carbon-intensive energy sources such as coal (Goldthau 2018; Raimi 2018; Bomberg 2017b).

Yet the economic benefits of fracking also came with myriad potential environmental costs. The extraction process triggered a vibrant scientific debate regarding the extent and probability of a range of environmental harms, including groundwater contamination, methane emissions, and earthquakes among others (Jackson et al. 2014; Small et al. 2014; Neville et al. 2017; Raimi 2018). Debates regarding scientific uncertainty spilled into the public sphere, where concerns about potential health and environmental harms generated widespread political resistance at local, subnational, and national levels (Davis 2012; Christopherson, Frickey, and Rightor 2013; Warner and Shapiro 2013; Fisk, Park, and Mahafza 2017; Wylie 2018). New coalitions of scientists, landowners, environmental groups, and community members mobilized in opposition to development (Weible and Heikkila 2016; Boudet et al. 2014; Neville and Weinthal 2016; Carter and Eaton 2016; Davis 2012).

At the heart of these conflicts was a classic tension in energy policy making in which policy makers had to assess the trade-offs between acquiring economic benefits and mitigating environmental harms. What was new in the case of hydraulic fracturing was the additional element of scientific uncertainty regarding technological risk. In the late 2000s, the rapid growth of production throughout the United States quickly outpaced the scientific community. The speed of technological innovation, the volatility of gas markets, and the paucity of environmental data generated substantial variability with regard to economic and environmental projections, including the size of resource plays, number of potential jobs, and depth of cumulative environmental harms (Sovacool 2014; Raimi 2018; Carter, Fraser, and Zalik 2017).

New drilling practices required experts in a variety of fields to assess the likelihood of hydrological, land-use, climate, and ecological habitat impacts, among others, all in the context of very little historical data. Indeed, an ongoing refrain in many scientific reviews was the absence of baseline data from which to assess probabilities of economic benefits and environmental harms (Rivard et al. 2014; Precht and Dempster 2012; Council of Canadian Academies 2014; Atherton et al. 2014; Cleary 2012; New Brunswick Commission on Hydraulic Fracturing 2016; Small et al. 2014; Sovacool 2014). As an emergent technology, hydraulic fracturing generated substantial scientific uncertainties regarding the assessment of potential economic and environmental costs and benefits.

In addition, hydraulic fracturing also generated significant political uncertainty due to intense and prolonged public concern, often concentrated in areas without a history of environmental mobilization (McAdam and Boudet 2012; Neville et al. 2017; Mazur 2016; Nikiforuk 2015; Neville and Weinthal 2016; Carter and Eaton 2016). Many subnational governments in the United States that were accustomed to low-salience technocratic energy policy making were caught off guard by the increasing prominence of the issue, a shift which in some cases prompted mobilized publics to push for increased engagement in local and subnational energy decision making (Davis and Fisk 2014; Rabe 2014; Davis and Hoffer 2012; Lowry 2008; Cleland et al. 2016; Cleland and Gattinger 2017).

Subnational governments in North America managed this uncertainty in different ways. Some US state governments subsumed hydraulic fracturing under their existing conventional oil and gas regulatory frameworks while others initiated temporary moratoria to facilitate further study (Rabe 2014; Richardson et al. 2013). Canadian provincial regulation has a distinctly regional character. To date, the western provinces of Canada have tended to develop single-issue approaches to regulation. Both the Campbell (2001–11) and Clark (2011–17) governments in British Columbia, as well as the Wall (2007–18) government in Saskatchewan, pursued substantial economic development, creating investment-friendly policy regimes. These governments have regulated environmental risks on an ad-hoc, single-issue basis.

In contrast, the Atlantic provinces experimented with a variety of regulatory approaches, even within a single provincial jurisdiction. In New Brunswick, the Alward (2010–14) government developed a comprehensive framework for hydraulic fracturing. The subsequent Gallant (2014–18) government announced a moratorium in December 2014, halting further development in New Brunswick. In Nova Scotia the Dexter (2009–13) government also implemented a moratorium in 2012. Newfoundland and Labrador followed suit by implementing a de facto moratorium in 2013, which was reaffirmed in 2016. Despite these broad regional divisions there were some outliers. In 2012 Alberta developed a more comprehensive regulatory framework than neighbouring Saskatchewan, while in 2014 Nova Scotia pursued a deeper commitment to precaution than its neighbours by announcing a legislated ban. This regulatory variation presents an empirical puzzle about the political dimensions of energy decision making: *(1) Why have provincial governments developed distinct regulatory frameworks for hydraulic fracturing? (2) How has policy change occurred in each province?*

A key determinant of hydraulic fracturing governance is the relative prioritization of economic benefits and environmental harms (Neville et al. 2017), a trade-off that is influenced by the interaction of structural conditions and contentious politics. Comparative public policy scholarship suggests that the influence of industry is likely a prime determinant of the shape of environmental regulation. Industry strength can be measured by the breadth and historical depth of production in each province, connected to the extent of the physical resource (Doern 2005; Carter, Fraser, and Zalik 2017). Western provincial economies have a long history of fossil fuel production, with stable policy subsystems dominated by industry representatives, bureaucrats, and scientists. Policy makers are highly attuned to the potential for hydraulic fracturing to provide an economic boost to local economies through job creation and economic development (Lindblom 1980; Carter, Fraser, and Zalik 2017). In this context, we would expect politicians' policy choices to be strongly determined by the market demands and the preferences of industry actors, leading to more lax environmental regulation (Hessing, Howlett, and Summerville 2005; Richardson et al. 2013; Carter 2020).

At the same time, research on US regulatory development suggests that environmental mobilization can play a mediating role on the influence of industry, leading toward more stringent regulatory frameworks (Rinfret, Cook, and Pautz 2014; Heikkila et al. 2014; Weible and Heikkila 2016; Fisk 2013; Richardson et al. 2013). Any impetus to limit the expansion of production lies in the strength of environmental mobilization to increase the salience of environmental harms among either policy makers or the electorate. To explain why and how variation in hydraulic fracturing regulation emerges in Canada, we need to examine the process through which provincial governments make the unlikely choice, namely how policy makers come to prioritize environmental harms over economic benefits under conditions of scientific and political uncertainty.

Risk Narratives

The analytical contribution of this book is to argue that the salience of environmental harms – and policy makers' corresponding prioritization of ecological costs – is mediated through the social construction of risk. Social science research on risk perception suggests that our evaluation of costs and benefits is substantially influenced by how we assess risk, namely how we determine the extent of hazards and the likelihood of their occurrence. This process is not objective but heavily mediated through our values, experiences, and social networks (Falkner and Jaspers 2012; Sunstein 2009). Amidst significant policy complexity and uncertainty, policy makers, like many of us, are "cognitive misers," and are unlikely to comprehensively assess the depth and extent of economic benefits and environmental costs based on scientific evidence (Boudet 2019). Instead, policy makers manage the constraints of time and limited resources by turning to ideas, heuristics, and worldviews, perspectives which are informed by perceptions of risk (Lindblom 1979; Simon 1985; Jones 1999; 2017; Boudet 2019). In short, how governments conceive of the risks of hydraulic fracturing shapes how they regulate the practice.

Governments' perceptions of risk are fuelled by collectively held beliefs within policy communities,[2] which I term *risk narratives*. Risk narratives justify and reinforce where regulatory development takes place, and who participates in policy formulation, two factors which strongly determine regulatory design (Pralle 2003; Baumgartner and Jones 1993; Scharpf 1989). How policy problems are defined not only shapes the policy communities' collective understandings regarding the dominant causal drivers of a particular policy challenge, but also limits the range of policy actions that are appropriate for government to take (Stone 1989; Majone 1989; Schon and Rein 1995; Hajer 1993; Baumgartner and Jones 1993; Pralle 2006a; Hall 1993; Blyth 2001; 2002). Problem definitions also set constraints on the appropriate venue in which policy development should take place and who should participate (Blyth 2013; Prindle 2012).

This book develops a typology of risk narratives to tease out relationships among problem definitions, processes of policy making, and regulatory outcomes. Drawing on findings from risk perception literature (Slovic 1987; 1993; 2000; Renn 2008; Pidgeon et al. 2017; Boudet 2019; Gastil et al. 2011), I argue that risk narratives can be measured along two dimensions: (1) unknown risk and (2) dread risk. Unknown risks refer to hazards which "are unobservable, unknown, new, and delayed in their manifestation of harm" (Slovic 1987, 283). Our perceptions of whether a phenomenon or technology is unknown is often fuelled by our ideas about scientific uncertainty, namely whether there is a relatively unified consensus among experts as to the relative extent of harms and a substantial historical record from which we can calculate probabilities – in the absence of both, radical uncertainty will be high (Blyth 2009).

In contrast, dread risks refer to hazards which have impacts that are involuntary, catastrophic, fatal, and inequitable (Slovic 1987). Perceptions of dread risk are driven by our sense of public urgency, which is often linked to the degree of catastrophic harm, together with our sense of fairness, which considers whether costs of a particular technology are equally distributed among groups or jurisdictions. My typology combines relative levels of unknown and dread risks to yield four ideal types: (1) linear, (2) complex, (3) uncertain, and (4) catastrophic. To tease out the relationships between narrative, policy making, and outcomes, I link each type of narrative with a decision making venue, which I distinguish as open or closed (Skogstad 2008; Howlett 2002), and lead actors who drive regulatory development, outlining a causal logic for both.

To address the question of *how* regulatory change occurs I also set out the ways in which risk narratives can facilitate different processes of learning within policy subsystems. Although studies have found that when individual respondents rank hazards as high on both dread and unknown risks they are more likely to want increased regulation (Slovic 1987; Gastil et al. 2011; Sunstein 2009; Boudet 2019), less is known about the relative trade-off between the two risk types. For example, how do policy makers respond when scientific uncertainty is high but public attention is low? Do they respond differently when

Figure 1.1. Analytical framework

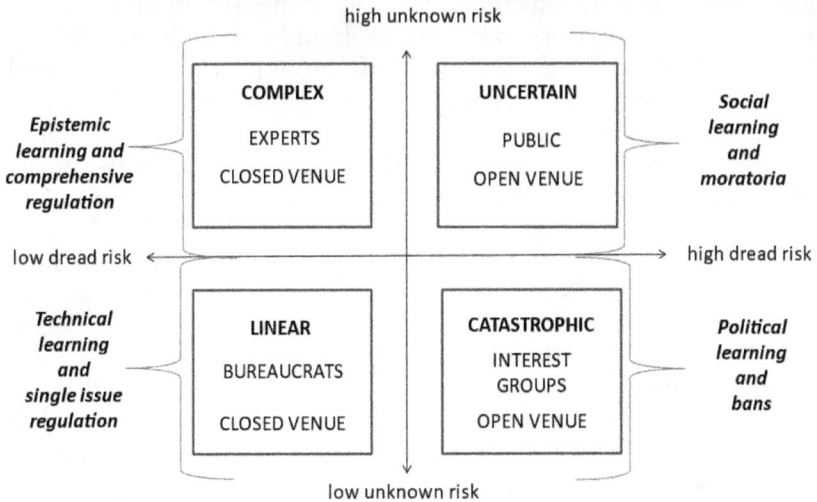

science is more certain, but the public is highly attuned to catastrophic harms? To answer these questions, I turn to policy studies literature on policy learning. Learning can be broadly understood as processes through which policy elites update their beliefs through "analysis, social interaction, rules and lived experiences" (Dunlop 2017, 215). Policy learning scholarship distinguishes four different modes of learning: (1) technical, (2) epistemic, (3) social, and (4) political (Rose 1991; May 1992; Hall 1993; Pierson 1993; Zito and Schout 2009; Heikkila and Gerlak 2013; Dunlop and Radaelli 2013; 2017; 2018a; 2018b; Boyd 2017; Moyson 2018). Theorizing suggests that shifts in problem uncertainty, and/or changes to who is a legitimate source of information can determine shifts in modes of learning (Dunlop and Radaelli 2013). Because risk narratives incorporate both problem uncertainty and legitimacy, I posit that these four different types of learning align with different types of risk narratives.

Figure 1.1 provides an overview of the analytical framework of the book, setting out four different ideal types of risk narratives and their relationship to dominant actors, venues, processes of policy learning, and subsequent regulatory outcome.

Linear risk narratives are those in which dread risks are minimal and uncertainties are low. As a result, linear risk narratives reinforce the notion that risks can be easily managed within relatively closed venues, namely existing government processes and line departments. In this case, the dominant actors in regulatory development are likely to be government officials and bureaucrats. Under these conditions, changes to regulatory frameworks are more likely to be at the level of instruments and settings, or single-issue regulation, as policy makers respond to low level harms under conditions of low issue salience. Linear risk narratives thus reinforce processes of *technical learning* in which policy elites, primarily bureaucrats, regulators, and elected officials update their beliefs about the efficacy of a particular policy instrument or setting.

Complex risk narratives are those in which unknown risks are high because of difficulties in assessing probabilities, but the potential for catastrophic or involuntary harm remains low. This narrative legitimizes policy makers turning to scientific expertise to develop regulatory designs and suggests that harms can be managed through state-academic collaboration. In this case, certain sets of experts, such as hydrologists, geologists, or economists, have a privileged role in the regulatory process. Depending on the science, changes could be at the level of settings, but they could also facilitate the creation of broader, more comprehensive regulatory frameworks (Renn 2008; Weible 2008). Complex risk narratives thus enable processes of *epistemic learning* in which government officials rely on input from academic experts, potentially updating beliefs about instruments and settings but also policy goals.

Uncertain risk narratives refer to those where unknown and dread risks are high. Uncertain narratives prompt governments to turn to scientific experts, but the dominance of scientific expertise can be tenuous and vulnerable to challenges to credibility (Cann and Raymond 2018). Uncertain narratives justify an opening up of institutional governance, creating opportunities for social learning through public consultations and external reviews (Millar, Davidson, and White 2020). Open decision-making processes can include a larger variety of participants, including those with knowledge of Indigenous perspectives and traditional knowledge, members of the public, journalists, and not-for-profit advocates. In this case, regulatory development has the potential to incorporate new policy directions or goals, reflecting a paradigm shift (Neville and Weinthal 2016; Skogstad 2003; Hartley and Skogstad 2005). Uncertain risk narratives thus facilitate processes of *social learning* where a broader collective of elites and publics update their deep core beliefs regarding the drivers of a policy problem.

Finally in *catastrophic risk narratives*, dread risks are severe and certain. This narrative legitimizes governments to act immediately and urgently protect the public interest. Under catastrophic risk narratives the driving actors in policy making are likely to be government officials and organized interest groups who wield the threat of broader public action. The venue for this type of regulatory development is likely to be in the public eye, but more adversarial, as in town halls or public demonstrations. The conditions are the most conducive to regulatory responses that prioritize precaution, such as bans. I suggest that catastrophic risk narratives are conducive to processes of *political learning* in which policy elites update their beliefs about the political feasibility of a given policy solution based on the positions of organized advocacy groups.

In sum, risk narratives generate political power for some sets of actors, while diminishing the clout of others in the policy process. As such, this research contributes to conversations in political science examining the ways in which policy ideas amplify or hinder the resources of different sets of actors within the policy-making process (Béland, Carstensen, and Seabrooke 2016). Risk narratives can strengthen the influence of different actors by bolstering their perceived credibility within the policy process. On balance, processes of technical and epistemic learning are more likely to lead to regulatory designs focused on single issues or clusters of regulations, while broader social processes are more likely to lead to moratoria or bans.

To probe the plausibility of the analytical framework this book traces regulatory development in British Columbia, Alberta, New Brunswick, and Nova Scotia from 2006 to 2016. These four English Canadian provinces have both demonstrated resource potential for shale gas development and embarked on some form of policy making and regulation of hydraulic fracturing. They were selected because regional comparisons allow me to leverage similarities in provincial culture, institutional structure, and political economy among the western provinces on the one hand and the Maritime provinces on the other. Paired comparisons help focus the analysis to explain fine-grained variation in regulatory frameworks as well as to interrogate the role of risk narratives, the ideational factor of interest. At the same time including both western and Maritime provinces allows for a country-wide comparison and situates this study within a burgeoning literature on subnational environmental comparative politics in Canada (Paquet and Broschek 2017; Boyd 2017; Rabe 2018; Carter, Fraser, and Zalik 2017; Carter 2020). I set a ten-year time frame to account for emerging policy development; however, the majority of regulatory changes examined in the book occurred between 2010 and 2014.[3]

To demonstrate the viability of the proposed analytical framework I draw on three main methodologies to inform the case studies:

(1) process tracing of energy regulation through primary and secondary document analysis; (2) thematic coding of risk narratives in provincial news media articles; and (3) key informant interviews with politicians, government officials, environmental advocates, industry representatives, grassroots activists, and scientific experts.

Overview

The next two chapters introduce the policy context and the analytical framework for the study. Chapter 2 examines the range of uncertainties at play in making policy on hydraulic fracturing, including technological, economic, and environmental concerns. I review the proposed economic benefits stemming from the practice and examine estimates of resource potential in Canada. To introduce the range of environmental harms facing policy makers during the analytic period I summarize risks identified in the grey and scientific literature, including relevant available information as to the degree of scientific uncertainty during the early 2010s. The chapter documents existing patterns in variation of regulation of hydraulic fracturing in the United States and traces a similar pattern in the Canadian context. This section also provides an overview of the timing of policy change in each province, demonstrating the temporal variation in decision-making processes at play in each jurisdiction.

Chapter 3 presents the analytical framework for the book. Drawing on comparative studies of fracking regulation, I review the ways in which structural conditions and contentious politics can affect regulatory design, hypothesizing that under conditions of uncertainty, risk narratives guide how policy communities assess the relative trade-offs between economic benefits and environmental harms. For each type of risk narrative (linear, complex, uncertain, catastrophic), I develop a series of proposals regarding who is legitimized to participate in policy formulation, where that process is located within policy-making institutions, and the type of learning each risk narrative is likely to trigger, along with predicted regulatory outcomes. The chapter concludes with an overview of the research design and methodologies used in the study.

Chapter 4 examines the case of British Columbia under the Campbell (2001–11) and Clark (2011–17) governments. I find that British Columbia was responsive to the needs of industry, a decision which was facilitated by low levels of unknown and dread risk. In media and policy debates a dominant frame of linear risk was put forward by industry representatives and government officials. This frame served to reinforce a process of technical learning, an approach that was facilitated by the delegation of regulatory authority to the BC Oil and Gas Commission. Findings suggest that the narrative of linear risk in media debates also served to ward off environmental opposition, a situation that was exacerbated by fragmentation within anti-hydraulic-fracturing coalitions.

Chapter 5 explores regulatory development in Alberta, focusing on decisions made by the Progressive Conservative Government under Redford (2011–14). I find that Alberta initially followed a process of technical learning driven by a linear risk narrative that was very similar to its western neighbour. However, the dominant narrative began to change as Albertan policy makers began to attend more closely to public opposition to hydraulic fracturing in the United States. The potential for strong public outcry based on perceived catastrophic environmental harms, together with the perceived failure of industry and government to "win" public debate regarding tar sands development, led to a narrative of more catastrophic risk than in British Columbia. Together with a contingent restructuring of the Energy Resources Conservation Board into the Alberta Energy Regulator, this risk narrative fostered a degree of political learning in which regulators developed a regulatory approach that anticipated public opposition, leading toward a more comprehensive regulatory framework.

Chapter 6 traces regulatory development in New Brunswick, examining the policy decision-making process of the Progressive Conservative government under Alward (2010–14) and Liberal government led by Gallant (2014–16). The study finds that, faced with what they perceived as unknown but relatively controllable risks, the Alward government turned to other regulators and scientific experts for new policy ideas to build public confidence. The process of epistemic learning spurred decision makers to believe in the capacity of the regulatory framework and, in particular, scientific expertise to build public trust. However, the newly developed scientific institutions failed to resolve or contain public conflict. Prior to the 2014 election, Liberal challenger Brian Gallant began to articulate an uncertain narrative in public debates. The uncertain risk narrative prompted processes of social learning in the province and almost immediately after gaining power in 2014 the government adopted a precautionary approach to manage political conflict, resulting in a moratorium, which it renewed in 2016.

Chapter 7 examines policy development in Nova Scotia, interrogating the policy making processes of the NDP government under Dexter (2009–13) and Liberal government under McNeil (2013–16). I find that faced with an initially low assessment of environmental harms, both on unknown and dread dimensions, the Dexter government engaged in a brief period of limited internal technical learning. But in response to dread risks generated by electoral pressures and localized mobilization the government soon engaged in processes of political learning, implementing a moratorium. The process of public consultation carried out by the subsequent McNeil government opened a venue for anti-fracking actors to put forward catastrophic risk narratives in public debates, which prompted another round of political learning in cabinet. Energy Minister Andrew Younger believed there was some scientific uncertainty regarding hydraulic fracturing, but, more important, he was attuned to the potential electoral risks of continued public mobilization both in Nova Scotia and in other jurisdictions. This process of political learning prompted Younger to implement a legislated ban.

Chapter 8 develops a comparison of findings from all four provinces, exploring the portability of the framework to other policy areas and practical implications for policy makers attempting to navigate the politics of risk in Canada and beyond. The book concludes with a brief epilogue exploring the emerging interpretive politics of natural gas in the context of Canadian climate action.

Chapter Two

Fracking Uncertainty

Political debates about hydraulic fracturing are debates about uncertainty. On the one hand there are uncertainties about economic benefits: What will the resource yield? Can the product reach domestic and international markets? Will gas replace coal in global markets? How many jobs will the new industry support? On the other hand there are uncertainties about environmental costs: What is the impact on surface, ground, and drinking water? What are impacts on air, land, and ecological habitats? Are there climate impacts? What are the health risks? As with many other emergent technologies, such as genetically modified organisms, carbon capture and storage, blue hydrogen, small modular reactors, smart grids, and wind turbines, many of these questions are contested, as science, law, and politics regarding the issues are nascent and underdeveloped (Boudet 2019; Falkner and Jaspers 2012). Within this swirl of uncertainties, policy makers attend to some costs and not others, taking up some narratives about relative risks and discarding others. Understanding the dimensions of these technological, economic, and environmental uncertainties sets the stage for examining provincial policy action.

Uncertain Technologies

At a basic level, even the notion that hydraulic fracturing is a *new* technology can be contested. Industry communications often stress that hydraulic fracturing has been used in conventional oil and gas production since the 1950s (Heffernan 2013; CCEI 2007). The use of injecting water to create cracks in oil and gas formations was indeed a fairly common practice in twentieth century conventional drilling. However, the innovation that transformed the industry and made unconventional production profitable has been the combination of high volume hydraulic fracturing with horizontal and directional drilling[1] (Raimi 2018; Zuckerman 2013).

To engage in modern hydraulic fracturing, a producer first has to build a well pad to drill a vertical well to substantial depths ranging from 4,000 to 11,000 feet (US EIA 2013b). At the foot of the vertical well a horizontal arm is drilled, often extending up to 3 km long (Council of Canadian Academies 2014, 45). The vertical well is often encased in two sets of casings: the surface casing, extending below the freshwater zone, and the intermediate casing, extending to just above the production zone. Once the well is drilled and the casings are cemented, the well is ready for hydraulic fracturing. Operators use high volumes of a fracturing fluid, which, dependent on the geology of the resource, can consist of water, gelled propane, or liquid carbon dioxide. Fracturing fluid is forced down the well at high pressures to generate fractures radiating out from the production casing. A mixture of chemicals and sand, which functions as a "proppant," ensures that the fractures remain open to allow for the release and capture of natural gas from the rock. One of the innovations that enabled significant yields in comparison to earlier iterations is that operators now use high volumes of fracturing fluid in multiple stages to generate fractures. Before production starts, flowback fluids that are often termed "produced water" are generated. Produced water tends to be a combination of the hydraulic fracturing liquids, water from the formation, and naturally occurring radioactive materials (NORMs), which must also be recovered and disposed of, either through surface discharge, recycling, or injection into deep saline aquifers through separate wells. In addition, any natural gas occurring with flowback fluids must be collected, vented, or flared (Raimi 2018).

The success of operators drilling in the Texas Barnett Shale play resulted in the rapid spread of hydraulic fracturing throughout the oil and gas industry in a relatively short period of time (Raimi 2018). The result was a substantial rise in hydraulic fracturing to a total of over 40,000 shale gas producing wells throughout the United States (Rivard et al. 2014). The US production of shale gas since the 1990s has been significant because of the speed of technological uptake by producers, reflecting rapid cycles of innovation. Geny (2010) documents the proliferation of hydraulic fracturing of horizontal wells in the Barnett Shale, noting a dramatic five-fold increase in just over two years. Figure 2.1 illustrates the spread of technology throughout the United States from 1996–2019.

This technological innovation was central to the growth of shale gas extraction in the United States, with the rising price of natural gas

Figure 2.1. Horizontal wells in the United States

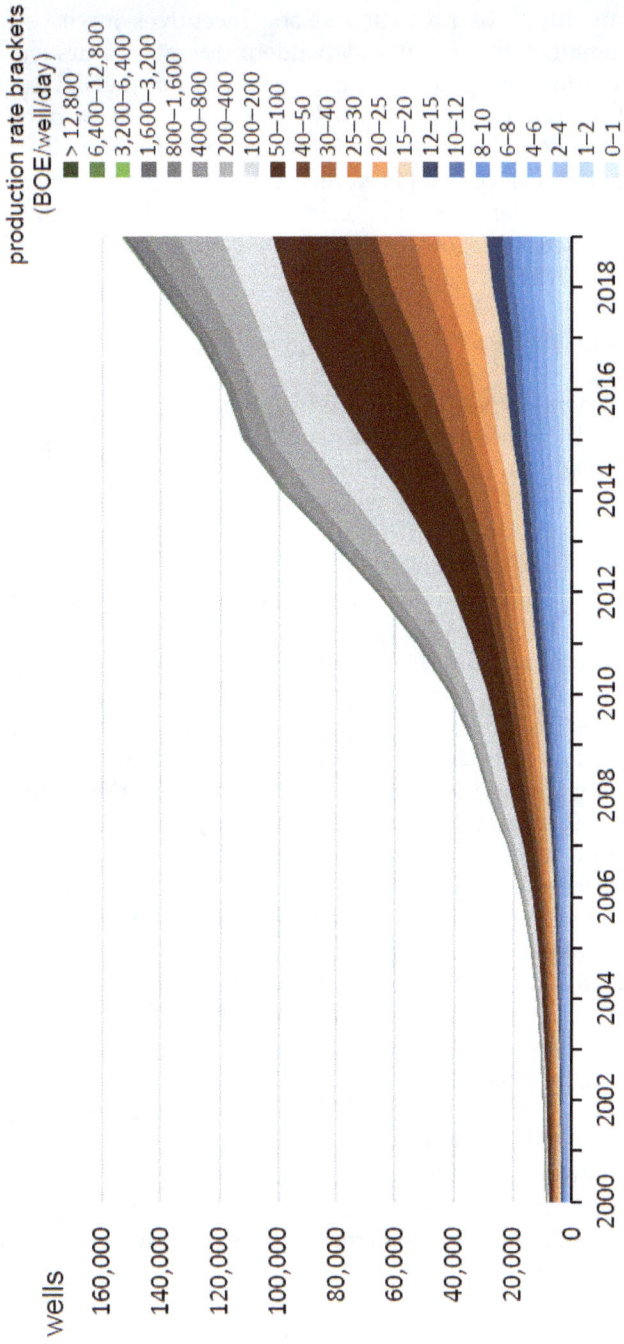

Source: US EIA (2020)
Note: BOE = barrel of oil equivalent

throughout the 2000s providing strong incentives for investment. A range of small, competitive, independent operators pursued production, generating competition and innovation for specialized services such as horizontal drilling, thus driving the spread of development (Geny 2010; McGowan 2014). Scholars have also credited federal research and development tax credits and subsidies, the deregulation of gas markets, and landowner subsurface rights as key policy drivers of US development (McGowan 2014; Geny 2010; Centner and O'Connell 2014; Davis 2012).

Uncertain Economics

The increase in US natural gas production due to hydraulic fracturing fundamentally transformed energy politics in North America and globally (Foss 2007; 2011; Raimi 2018; Goldthau 2018; Neville et al. 2017). Estimates from the US Energy Information Administration demonstrate that US production for both dry shale gas and natural gas liquids pivoted from declining yields in the early 2000s to unprecedented high levels of production in 2018, levels which are projected to be stable through to 2050 (US EIA 2019). This increase reflects a more than threefold increase over year 2000 production levels (US EIA 2014; 2013a; 2017).

The domestic production of unconventional gas in United States has fundamentally changed the Canadian-American relationship, transforming Canada's position as a preferred American importer of natural gas (Gattinger 2015; 2012). The United States, historically an importer of Canadian natural gas and crude oil, became an exporter of liquid natural gas (LNG) in 2017 (US EIA 2019). The US EIA predicts an ongoing decline of western Canadian imports and an increase in US exports to eastern Canada throughout the 2020s (US EIA 2014; 2017). This reduction in US demand has forced Canadian suppliers to seek out new Asian markets (Pembina Institute 2012; BCBC 2013; Mirski and Coad 2013). Global markets are shifting to more intense supply-side competition, as producers aim to leverage the high prices of gas tagged to oil in the Japan/South Korea/Taiwan markets and the low costs of production in North America. Long term predictions anticipate a movement away from gas tagged to oil towards more spot pricing based on North American hubs (Foss 2011; Ernst & Young 2013). Canada's main competitors in accessing Asian markets are likely to be either Australia or the United States. Both countries are further ahead in either greenfield construction or the conversion of existing LNG import facilities to export terminals (Ernst & Young 2013).

The rapid increase in US natural gas production from 2006 on has resulted in significant volatility in North American domestic natural gas prices, from a high of $13 per million Btu (MMBtu)[2] in 2008 to a low of $1.80/MMBtu in 2009 (Foss 2011). The US EIA predicts that the domestic spot price of natural gas at Henry Hub[3] will level out at $4/MMBtu from 2022 to 2050 (US EIA 2022).

In addition to price uncertainty, resource estimates of the extent of recoverable shale gas and tight oil have been variable because of technological and statistical difficulties in applying conventional estimation practices (Lavoie et al. 2012). Nevertheless, estimates have identified over 688 shale basins worldwide, including both shale gas and shale oil (Geny 2010). The main application of hydraulic fracturing to date has been with regard to targeting natural gas in shale plays,[4] but the technology has also been applied to "tight oil" production from shale and nearby sandstone, limestone, or siltstone plays, with significant production in North Dakota and Montana (NEB 2011). In 2013 the US EIA documented current technically recoverable global resources at 340 billion barrels of world shale oil resources and 6,964 trillion cubic feet (Tcf) of world shale gas resources (US EIA 2013b; 2015). Tables 2.1 and 2.2 identify the top ten countries with technically recoverable shale gas and oil resources.

As evident in figure 2.2, the geological formations of shale gas and oil are expansive throughout North America, with heavy concentrations in the United States and Western Canada. In 2021 US shale gas production totaled 24 trillion cubic feet, with the majority of production concentrated in the Southwest and Eastern regions (US EIA 2022).

To date most shale gas development in Canada has been exploratory. Nevertheless, the potential for industry expansion is significant; shale gas and tight oil plays have been identified in British Columbia, Alberta, Saskatchewan, Manitoba, Ontario, Quebec, and the Atlantic Provinces as well as in Nunavut, the Yukon and the Northwest Territories (NEB 2009; Natural Resources Canada 2015). Estimates for the Horn River Basin, one of two major shale plays located in northeastern British Columbia and Northwestern Alberta, calculated a case for marketable natural gas between 78 and 133 trillion cubic feet (Tcf) (NEB and BCMEM 2011; US EIA 2015).[5] In 2011 there were approximately 855 gas wells operating in the Horn River Basin, Montney play, Liard Basin, and Cordova Embayment (C. Adams 2011). In 2013 500 wells were drilled to target unconventional production, representing 85 per cent of total wells drilled (BC Oil and Gas Commission 2014b). From 2016 to 2020 the average annual number of natural gas completions in the province has been 380 wells per year (CAPP 2022). Production in

Table 2.1. Top ten countries with technically recoverable shale gas resources

Rank	Country	Shale Gas (Tcf)
1	China	1,115
2	Argentina	802
3	Algeria	707
4	United States	665
5	**Canada**	**573**
6	Mexico	545
7	Australia	437
8	South Africa	390
9	Russia	285
10	Brazil	245

Source: US EIA (2013b)

Table 2.2. Top ten countries with technically recoverable shale oil resources

Rank	Country	Shale oil (billion barrels)
1	Russia	75
2	United States	58
3	China	32
4	Argentina	27
5	Libya	26
6	Australia	18
7	Venezuela	13
8	Mexico	13
9	Pakistan	9
10	**Canada**	**9**

Source: US EIA (2013b)

Alberta has been less developed, but geological studies estimate marketable resources from four different formations to be 200 Tcf of natural gas (US EIA 2013b).[6] Alberta and Saskatchewan also have significant capacity for shale oil development, with US EIA estimates at 7,240 and 1,600 million bbl respectively (US EIA 2013b). Estimates of technically recoverable resources for the Utica shale in Quebec and the Nova Scotia Horton Bluff shale are significantly lower, at 31.1 and 3.4 Tcf respectively (US EIA 2013b). Technically recoverable estimates for the Frederick Brook shale in New Brunswick were not developed, but gas in place estimates identified approximately 75 Tcf (Rivard et al. 2014).[7]

Figure 2.2. Global shale gas and oil resources

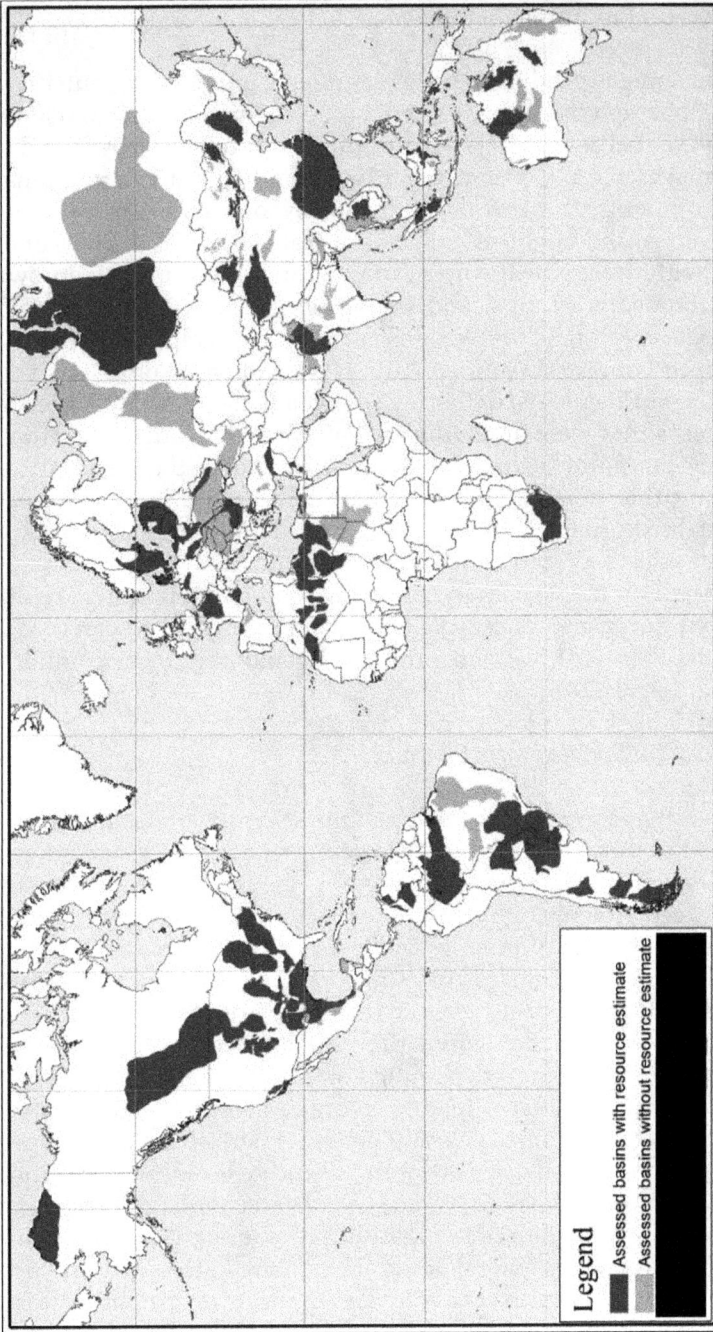

Legend

Assessed basins with resource estimate

Assessed basins without resource estimate

Source: US EIA (2013b)

Uncertain Environmental Science

As oil and gas producers have identified exploitable resources across the globe, environmental scientists have stressed that fracking has the potential to generate significant health-related harms and environmental risks. Concern regarding drinking water contamination, air pollutants, the loss of freshwater resources, increased methane emissions, and land use impacts generated varying degrees of attention within the North American scientific community, land owners, environmental groups, and the US electorate (Rabe 2014; Jackson et al. 2014; Jacquet 2014; Vengosh et al. 2014; Small et al. 2014; Arnold, Nguyen Long, and Gottlieb 2017; Hays and Shonkoff 2016). The following outlines some of the key environmental, health, and economic concerns that were emerging in the US and Canadian scientific and policy literature during the analytic period of the book (2006–16). During this time period scientific literature on potential environmental risks in the US context was rapidly developing and outpaced social science on political and social dynamics (Neville et al. 2017). Research on the Canadian context was much more sparse with provincial governments only beginning to engage in research collaborations with national and provincial geological agencies after 2010 (Rivard et al. 2014).

Ground and Surface Water Protection

One of the key concerns arising in the scientific literature during the period of study was limited knowledge regarding the hydrology and geology of the "intermediate zone" – the layer of strata separating the "deep" fracturing site and the fresh groundwater zone that functions as a source of potable water (Council of Canadian Academies 2014, 68; Vengosh et al. 2014). Areas of concern included the possibility that fractures could extend vertically from the production zone, creating channels that would allow methane to flow up and contaminate freshwater aquifers (Council of Canadian Academies 2014, 68). Industry analysts have tended to argue that the probability of this type of contamination is low (Pembina Institute 2012). However, in their comprehensive review of the scientific literature the Council of Canadian Academies (2014) argued that scientific data are limited, noting that sufficient data regarding contamination and cumulative effects have not been collected. The council stressed that "a claim that shale gas developments have no impacts on groundwater needs to be based on generally accepted science including appropriate data obtained from the groundwater system using modern investigative methods" (68). Policy analysts and government officials have formally acknowledged a significant lack of baseline scientific information regarding groundwater quality, characteristics, volume, and movement in Canada (Pembina Institute 2012; K. Campbell and Horne 2011; Rivard et al. 2012; Parfitt 2011; Energy and Mines Minister's Conference 2013b).

The Geological Survey of Canada as well as government consultants have encouraged baseline testing of water quality prior to drilling (Rivard et al. 2012; 2014; Precht and Dempster 2012; ERCB 2011). For example, Geoscience BC partnered with Encana and other producers to characterize the watersheds in the Horn River Basin (Salas and Murray 2013). Beyond recommending further study, regulators in the United States and Canada also quantified minimum setback distances between wells, as well as between the production string and freshwater aquifers (Precht and Dempster 2012; Richardson et al. 2013), while some environmental NGOs advocated for moratoriums to be established prior to acquiring additional hydrological data (Parfitt 2011).

Drilling and Well Construction Failures

Similar to risks with conventional oil and gas development, a significant environmental challenge with regard to shale gas development is the potential for faulty well construction to result in contamination of freshwater aquifers, venting of methane, or contamination of other drinking water wells in close proximity to the drill site (Council of Canadian Academies 2014; Parfitt 2011; BC Oil and Gas Commission 2012a). Potential pathways to contamination include faulty cement in either the surface or intermediate casings or failure to adequately cement the well casing to the surface, as well as leakage from abandoned wells (Council of Canadian Academies 2014). Studies suggested that the probability of blowouts is higher for shale fracturing than in conventional gas drilling, with two reported cases in Western Canada in the late 2000s/early 2010s (Pembina Institute 2012, 28). Industry solutions were aligned with conventional drilling practices, including requiring mandatory surface casings, using recommended intermediate and production casings, and checking cement quality using methods such as bond logs, pressure testing, or plugs (Council of Canadian Academies 2014; Precht and Dempster 2012; Canada 2013).

Water Management

Given that the process of hydraulic fracturing tends to require extensive volumes of water, identified risks included the availability of water sources, management of required volumes, and the possibility of conflicts between the oil and gas industry needs and other users, especially municipalities. Potential sources for water are from "surface basins, shallow ground water (which is typically fresh), deep groundwater (which is typically saline), and or/recycled water (either flowback water or municipal waste water)" (Pembina Institute 2012, 25). Water withdrawals to date in Canada have generally been from the surface, although producers in the Horn River have used ground water (Pembina Institute 2012). Estimates of projected water use are extremely variable; depending on the geology of the shale formation, they range from 2,000–35,000 cubic metres per well (Johnson and Johnson 2012).[8] Water use in Canada is highest in the Horn River Basin; Geoscience BC reported 7 million cubic metres of water were used in 133 Horn River oil and gas wells in 2011 (Salas and Murray 2013). Table 2.3 provides an overview of projected annual water volume required for wells in different formations in western Canada.

Environmental and Indigenous groups have voiced concerns regarding the adequacy of consultation regarding water withdrawals and the lack of coordination between the BC Oil and Gas Commission and the Ministry of Forests, Lands, and Natural Resource operations regarding short- and long-term leases for surface water withdrawals (Parfitt 2011). Potential policies to manage water consumption include new water licensing legislation, either including caps or thresholds (Precht and Dempster 2012) or incentivizing the use of saline or brackish water sourced from deep aquifers (Pembina Institute 2012; K. Campbell and Horne 2011). Encana and Apache experimented in the early 2010s in BC with deep saline withdrawal and re-injection (Canada 2013). However the use of saline water also poses problems for fracturing in terms of additional corrosion (Heffernan 2013). Other options include recycling produced water for re-use in secondary fractures; this option has also been piloted in northeastern British Columbia by Shell (Pembina Institute 2012; Parfitt 2011). More assertive policy options include developing a water pricing market that more accurately reflects the externalities of unconventional oil and gas development (Parfitt 2011, 35).

Water Quality

Beyond the scale of water use, one of the more salient risks identified in the literature is the use of chemical additives in fracturing fluids.

Table 2.3. Projected water volume use in western Canada shale formations

Formation	Resource	Volume per well in cubic metres
Horn River (BC)	Natural gas	35,000–135,000
Montney (BC, AB)	Natural gas, liquids, oil	2,000–15,000
Duvernay (AB)	Natural gas, liquids	10,000–60,000
Cardium (AB)	Oil	0–4,000
Bakken (SK)	Oil	0–3,000

Source: Heffernan (2013)

The process requires a range of chemical components, including acids, friction reducers, surfactants, salts, scale inhibitors, pH adjusting agents, iron control, corrosion inhibitors, and biocides (Pembina Institute 2012). While these additives usually comprise less than 1% of the total mix of fracturing fluid (Council of Canadian Academies 2014), the high volumes of fluid required mean that the absolute quantities of substances can be significant (Pembina Institute 2012; Council of Canadian Academies 2014). Documented spills and leakages of fracturing fluids in the United States, often from storage tanks, have raised concerns regarding the release of carcinogens into the environment, particularly during transportation and storage (Pembina Institute 2012; Vengosh et al. 2014).

Mandatory disclosure of chemicals on public websites[9] has emerged as an industry standard to respond to public concern, although the effectiveness of this practice has been subject to several critiques in the US context, particularly with regard to the ability of companies to withhold data in order to protect proprietary data or "trade secrets" (Precht and Dempster 2012; Fisk 2013; Dundon, Abkowitz, and Camp 2015; Konschnik and Dayalu 2016). Observers in the Canadian context have identified several gaps in federal/provincial coordination regarding categorization of chemical substances, stressing that oil and gas companies are exempt from the federal National Pollutant Release Inventory (Pembina Institute 2012; Parfitt 2011). In 2012 the Canadian federal Commissioner of the Environment and Sustainable Development highlighted that federal departments had yet to carry out environmental risk assessments for the majority of chemicals commonly used in hydraulic fracturing (OAGC 2012). Other alternatives include the development of "green" additives (Rivard et al. 2012) or non-toxic tracers to document the migration of fracturing fluid (Canada 2013).

Water Disposal

Linked to the discussion of chemical disclosure is the concern that produced waters also contain a variety of substances or chemicals that could pose substantial environmental and health risks if disposed of ineffectively. Produced water can contain heavy solids and naturally occurring radioactive materials (NORMs), which could be harmful if discharged to surface water through inadequate water treatment facilities, leaks, or spills (Council of Canadian Academies 2014; Lauer, Harkness, and Vengosh 2016, 5389). To date, treatment of produced water has been challenging because the high levels of solids make it less likely for water to be recycled in existing municipal treatment facilities under current economic conditions (Pembina Institute 2012; Jackson et al. 2014). In the United States, producers have used a variety of earthen, open, lined pits, or closed tanks to contain produced water prior to transportation (Richardson et al. 2013). Deep well injection of produced water, which requires the construction of new disposal wells or repurposing of older conventional wells, is common in Alberta gas production and is increasingly being used as an option in British Columbia (Pembina Institute 2012). Unlike some US states, surface discharge of water is not permitted in most Canadian jurisdictions (Pembina Institute 2012).

Seismic Activity

One risk which has gained significant attention in both Canada and the United States is the link between shale gas development and seismic activity, or earthquakes (Walsh 2014). Since 2008 there has been a documented rise in seismicity throughout the United States, which has been linked to the downhole (deep) injection of produced waters into disposal wells (Pembina Institute 2012; BC Oil and Gas Commission 2012a; 2014c; Andrews and Holland 2015; Allison and Mandler 2018; Schultz et al. 2018). The specific process of hydraulic fracturing has been less likely than deep injection water disposal to trigger earthquakes of significant (felt) magnitude, although there are documented exceptions, including the Canadian Duvernay shale play (Allison and Mandler 2018; Schultz et al. 2018). Although the causal link has been identified by a range of scientists and industry operators, regulatory options are less developed, with analysts recognizing that the issue tends to be under-regulated, especially in regions where seismic activity is previously unheard of (Precht and Dempster 2012; Rivard et al. 2012). Since 2015 state regulators in the United States have begun to experiment with limiting the injection rates and/or volumes of water disposal and specifying restrictions on the location of disposal wells (Allison and Mandler 2018).

Land Use and Wildlife Impacts

Perspectives in the literature regarding land use impacts are somewhat contradictory. The oil and gas industry tends to argue that land impacts are more limited than with conventional gas. Multiple horizontal wells can be drilled from the same well pad, reducing the need for the numerous well pads common with conventional vertical drilling (Heffernan 2013). However, the geology of the resource likely requires increased subsurface well density to make drilling economically viable (ERCB 2011). The latter raises the possibility that well pads will be larger, resulting in a larger land use footprint (Pembina Institute 2012; Council of Canadian Academies 2014). Production also requires a developed road infrastructure to truck in high volumes of water and sand as well as transmission pipelines to transport the produced gas and liquids to markets (Pembina Institute 2012). Assessment of wildlife impacts to date suggests that infrastructure development likely leads to habitat fragmentation, increased wildlife mortality, changes in streams and loss of aquatic habitat, and loss of species and vegetation (Council of Canadian Academies 2014; Olive 2018). Reports from livestock owners also suggest that exposure to produced water (from spills and improper surface storage) has had significant impacts on reproductive cycles of cattle and pets (Bamberger and Oswald 2016). Overall, the lack of consensus on probabilities of groundwater contamination, whether from chemical migration, spills, faulty construction, and wastewater disposal methods, makes it difficult to assess the impacts of hydraulic fracturing on plants and animals, as well as estimating cumulative effects (Souther et al. 2014).

Climate

One of the more contentious issues within the policy and scientific literature is the implication of shale gas development for climate change mitigation and GHG reduction goals (Stephenson, Doukas, and Shaw 2012). Hydraulic fracturing proponents promote unconventional gas as a potential transitional energy source in shifting North American markets to a low-carbon economy, arguing that unconventional gas is a cleaner alternative than coal while still more cost effective than renewable energy (Delborne et al. 2020). Critics argue that shale gas will stall deep decarbonization by competing with renewable energy sources, thereby locking economies into higher emissions trajectories (Hazboun and Boudet 2021; Delborne et al. 2020; Chen 2020b; Brauers 2022). Substantial academic controversy and debate exists regarding underlying assumptions in GHG lifecycle modelling of shale gas development (Howarth, Santoro, and Ingraffea 2011; 2012; Howarth 2014; Cathles et al. 2012). GHG emissions are generated by fluid transportation, use, and disposal; the length/extent of drilling; the venting of gas flowback; and the formation of CO_2 (Pembina Institute 2012; Goehner and Horne 2014; Jackson et al. 2014). Jaccard and Griffin (2010) find concentrations of CO_2 in the Horn River Basin gas of approximately 10%, which is two to three times more than the average CO_2 content of conventional natural gas reservoirs (4). Current regulation for conventional gas production aims to reduce venting by incentivizing flaring. However the stringency of the regulation is often mitigated by economic qualifiers; e.g., venting is permitted if it is not economically feasible to conserve or flare gas (Pembina Institute 2012; Richardson et al. 2013). Other suggested options include mandatory carbon capture and storage for gas processing plants (Jaccard and Griffin 2010; Goehner and Horne 2014), caps on natural gas production (Parfitt 2011), taxes on methane emissions (Small et al. 2014), and "no-go zones" to limit the overall scale of industry development (Parfitt 2011).

More broadly, shale gas production from fracking also has significant implications for global decarbonization pathways. Projections by the International Energy Agency (IEA 2019) stress that natural gas has the potential to facilitate transitions to lower-carbon economies by replacing more carbon-intensive energy sources such as coal. On the other hand, the production of natural gas can also serve to reinforce carbon lock-in by slowing the advancement of renewables such as wind and solar PV (International Energy Agency (IEA) 2021; Sovacool et al. 2020).

Responding to Uncertainties: Subnational Regulation

How policy makers and regulators responded to the promise of economic benefits and the potential for environmental harms varied substantially across regions. US state regulation is highly fragmented and decentralized with variation between states (Boersma and Johnson 2012; Warner and Shapiro 2013; Rabe 2014; Small et al. 2014; Zirogiannis et al. 2016; Schenk et al. 2014). By and large, the locus of power for regulation is situated with state governments, although some municipalities in some jurisdictions have sought regulatory powers (Rabe and Borick 2013; Christopherson, Frickey, and Rightor 2013; Arnold and Holahan

Table 2.4. Matrix of regulatory tools for shale gas

Elements/activities	Regulatory tools
Site selection and preparation	General well spacing rules
	Building setback requirements
	Water setback requirements
	Predrilling water well testing requirements
Drilling the well	Casing/cementing depth regulations
	Cement type regulations
	Surface casing cement circulation rules
	Intermediate casing cement circulation rules
	Production casing cement circulation rules
Hydraulic fracturing	Water withdrawal limits
	Fracturing fluid disclosure requirements
Wastewater storage and disposal	Fluid storage options
	Pit liner requirements
	Underground injection regulations
	Fluid disposal options
Excess gas disposal	Venting regulations
	Flaring regulations
Production	Royalties
Plugging and abandonment	Well idle time limits
	Temporary abandonment limits
Other	State and local bans and moratoria
	Number of regulatory agencies

Source: Adapted from Richardson et al. (2013: 9)

2014; Arnold and Neupane 2017; Baka et al. 2018). State regulators have also been reluctant to cede power to the federal government, in part because of the strength of the industry's preference for state-level regulation (Baka et al. 2018; Rabe 2022). Table 2.4 provides a summary of the different types of regulation in place in the United States pertinent to practices of hydraulic fracturing and horizontal drilling.

Building on earlier case study work in Colorado, Philadelphia, Texas, and Illinois (Rabe and Borick 2013; Davis 2012; Davis and Hoffer 2012; Rahm 2011), Rabe (2014) argues that early regulatory regimes suggest an incremental trend towards pro-development, with the majority of state governments focusing on single-issue legislation, such as chemical disclosure or disposal of produced waters. Rabe also finds that some states have passed more comprehensive legislation, demonstrating a significant variation between either a pro-development policy regime ("race to the bottom") or a more extensive environmental regulatory model ("race to the top"). Rabe finds little support for policy coherence through mechanisms of diffusion. He argues that the inheritance of existing oil and gas regulation, together with different frameworks for water regulation, has impeded policy coherence or harmonization across states. The heterogeneity in US state regulation suggests that regional or subnational contexts are important in determining policy outcomes, highlighting the need for cross-case comparative studies in the Canadian context.[10]

Similar to the patterns identified by Rabe (2014) in the United States, some provincial governments in Canada have also adopted a single-issue approach to regulation of hydraulic fracturing, others have developed more comprehensive frameworks, and some have issued temporary moratoriums (Carroll, Stephenson, and Shaw 2012; Stephenson and Shaw 2013; Lachapelle, Montpetit, and Gauvin 2014; Montpetit, Lachapelle, and Harvey 2016; Montpetit and Lachapelle 2017; Carter and Eaton 2016; Gagnon et al. 2015; Neville and Weinthal 2016; Olive 2016; Ingelson and Hunter 2014; Olive and Delshad 2017; Garvie and Shaw 2014; Larkin et al. 2018).[11] As of the end of 2016, Quebec, New Brunswick, Newfoundland and Labrador, and the Yukon had implemented formal or de facto moratoria on the practice of hydraulic fracturing and Nova Scotia had implemented a ban on high-volume hydraulic fracturing (Cousineau, Marotte, and Seguin 2012; McHardie 2016; Government of Newfoundland and Labrador 2016; CBC News 2015; Erskine 2014). As a result, there are no specific regulatory standards in place in these provinces/territories. Table 5 provides an overview of the regulatory tools in place as of 2016 in British Columbia, Alberta, and Saskatchewan, as well as the 2013 rules enacted in New Brunswick.[12] The table compares these tools with standards established by the American Petroleum Institute as per Richardson et al. (2013).

As table 2.5 illustrates, provincial regulation is most mature in construction standards such as specifying drill casing and cement testing requirements. Regulations relating to withdrawals, treatment, and storage are less developed. By and large provinces have followed the lead of US state regulators in using the website fracfocus.ca as a guide to manage chemical disclosure, which provides a consistent, albeit limited baseline of reporting (Dundon, Abkowitz, and Camp 2015; Konschnik and Dayalu 2016). Provinces vary in the degree to which they have established prescriptive quantitative standards for well siting. Although New Brunswick's 2013 guidelines establish specific distances for setbacks, Alberta and British Columbia both moved away from spacing regulations in 2011 in order to increase the efficiency of proponents in targeting shale resources (confidential interviewee 2015c; 2015b). In

Table 2.5. Overview of Canadian regulation of hydraulic fracturing

Activities	Regulatory Tools	BC	AB	NB	SK	API
Overall policy		No	Yes	Yes	No	N/A
Site selection and preparation	General well spacing rules	Subsurface spacing deregulated in 2012	Subsurface spacing deregulated in 2011	No standard	1 section or 259 hectares	No standard
	Building setback requirements	100m	100m	500m school, hospital, nursing home; 250m dwelling, 250m outdoor public concourse, 100m permanent building, railway, road	100m water body; occupied dwelling; public facility or urban centre	Discretionary
	Water setback requirements	200m water well 100m surface water	200m horizontally, 100m vertically	500m public water; 250 communal water; 250 surface intake; 250 surface water supply	180m separation between water wells and seismic activity	Discretionary
	Predrilling water well testing requirements	Pre-fracture water testing of water wells within 200m of drill site if agreed to by well owners; chemical testing case by case	AE&W Standard for Baseline Water-Well Testing for Coalbed Methane/Natural Gas in Coal Operations	Baseline testing of water wells within 500m of drill site	Not required	Discretionary

(Continued)

Table 2.5. (Continued)

Activities	Regulatory Tools	BC	AB	NB	SK	API
Drilling the well	Casing/cementing depth regulations	Performance standard cemented to surface	25m below free gas zone surface casing cemented to surface; intermediate cemented 100m above geologic zone	25m below free gas zone. Based on AB directive –25m below sand and gravel	Casing required minimum depth of 20m below glacial drift; 10% of total well or 75m	30.5m (100 feet) below freshwater zone
	Cement type/ testing regulations	Logging required prior to abandonment	Logging required prior to abandonment	Logging required prior to drilling	Pressure tested at surface 8 hours before drilling. Logging not required.	
	Casing strings	Triple casing	Double or triple casing	Triple casing	Double casing	Triple casing
Hydraulic fracturing	Water withdrawal limits	Groundwater withdrawal greater than 75L/s requires an environmental assessment	Temporary and permanent licences more than 5000m on Crown lands requires additional approval; no approval needed for saline	More than 50m/ day triggers environmental impact assessment	Surface withdrawals require licence from water authority; plan re groundwater must be reported	Recommends consultation
	Fracturing fluid disclosure requirements	Mandatory disclosure via frac focus website	Mandatory disclosure via frac focus website	Mandatory via EIA process; mandatory disclosure via frac focus website	Voluntary disclosure via frac focus website	Recommends operators disclose fluid composition upon request

Activities	Regulatory Tools	BC	AB	NB	SK	API
Wastewater storage and disposal	Fluid storage options	Open, closed or earthenware tanks	Synthetically lined storage systems	No earthen pits – tanks and closed containers required	Tanks	Open pits
	Pit liner requirements	Lined	Lined	TBD	Not required	Discretionary
	Underground injection regulations	Permitted under Oil and Gas Act	Injection wells regulated under Directive 51	Not current practice due to geology	Approval required	Recommends injection where possible
	Fluid disposal options	No surface discharge; encourages reuse and recycling	No surface discharge	Recycling; shipped to NS and QC for disposal	Flowback to be treated	API encourages recycling
Excess gas disposal	Flaring and/or venting regulations	Guidelines to reduce venting; 2016 Climate Leadership Plan target a 45% reduction in vented emissions by 2025.	Venting allowed only if gas conservation has been determined to be economically infeasible and flows will not support stable combustion (Directive 060)	No venting from wellbore; must be flared or captured		API encourages flaring of all gas not captured for sale

Sources: Adapted from (Precht and Dempster 2012; Council of Canadian Academies 2014; Richardson et al. 2013; Government of New Brunswick 2013a; Government of Newfoundland and Labrador 2013; ERCB 2011; Government of British Columbia 2007a; 2016). Note that the NB specifications were rules for industry, not formal regulations

Figure 2.3. Canadian regulatory variation

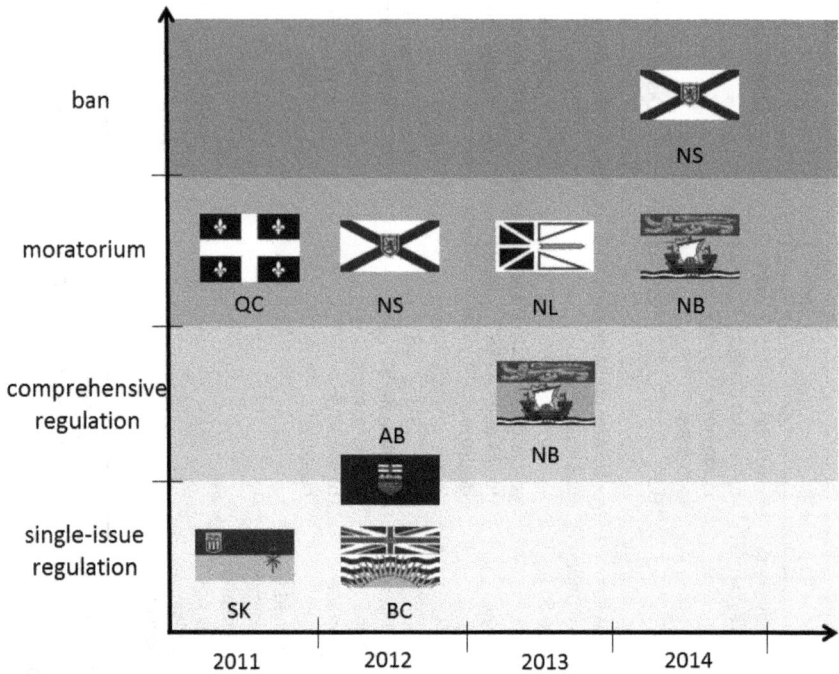

sum, New Brunswick's 2013 rules present the most comprehensive regulatory framework while Saskatchewan's regulatory framework is the least tailored to hydraulic fracturing.

Figure 2.3 places provincial policy and regulatory frameworks for hydraulic fracturing along a continuum, falling into four broad categories: (1) single-issue regulation; (2) comprehensive regulation; (3) moratoria; (4) bans. Single-issue regulation refers to regulatory frameworks that have focussed on a particular aspect of hydraulic fracturing – for example chemical disclosure rules. Comprehensive regulation refers to a framework that is tailored to hydraulic fracturing and regulates across a broad scope of elements and activities. Moratoria include formal and de facto "holds" on production; the assumption here is that although they can be enforced over the long term, different moratoria include exceptions for research or exploration and can be easily reversed by the executive. Legislated bans are presumed to be more durable, or at least administratively burdensome to reverse, and are more likely to signal that the jurisdiction is "closed for business." Figure 2.3 categorizes major policy/regulatory changes in each province from 2011–2014.[13]

Initially British Columbia, Alberta, and Saskatchewan adapted settings of their existing oil and gas regulatory regimes, with very minor, if any, adjustments to conventional oil and gas regulation. Nevertheless, although the three provinces seem relatively harmonized, particularly regarding drilling, there are some exceptions. In 2012, the Alberta regulator (at the time the Energy Resources Conservation Board) developed a specific hydraulic fracturing discussion paper that suggested a movement towards a more comprehensive policy response focusing on play-based regulation (ERCB 2012a). Although British Columbia did not develop a comprehensive regulatory regime specifically for hydraulic fracturing, in 2012 the government announced that it would make chemical disclosure of fracture fluids mandatory. Their LNG strategy (Government of British Columbia 2012b) focused on shale gas development in northeastern British Columbia, building on earlier development initiatives outlined in the 2002 and 2007 Energy Plans and the 2012 Natural Gas Strategy (Government of British Columbia 2002; 2012a; 2007b).

In comparison, New Brunswick initially developed a comprehensive regulatory framework especially targeted at hydraulic fracturing, evident in its Rules for Industry and The Oil and Gas Blueprint (Government of New Brunswick 2013a; 2013b), although this was replaced by a moratorium introduced by a new government in 2014.[14] Newfoundland and Quebec also issued moratoria in the early 2010s, which were reaffirmed in 2016 following independent reviews (Government of Newfoundland and Labrador 2013; 2016; Rivard et al. 2014). Nova Scotia shifted from declaring a de facto moratorium in 2012, to commissioning an external review in 2013, to introducing a ban on high-volume hydraulic fracturing in the fall of 2014 (Government of Nova Scotia 2012a; 2014a; Erskine 2014). Ontario has not issued any exploratory licences for hydraulic fracturing although it commissioned a report on regulation in other jurisdictions and a gap analysis in 2013 (Energy and Mines Minister's Conference 2013b) and indicated that if it were to engage in hydraulic fracturing, regulatory changes would have to be made (Natural Resources Canada 2017).

Regulatory analysis thus demonstrates that British Columbia and Saskatchewan have focused on incremental development of single-issue settings, in keeping with trends identified in Texas (Davis 2012). New Brunswick's initial comprehensive regulatory approach was reflective of either Illinois or Pennsylvania (Rabe 2014), although the comprehensive framework was replaced by a moratorium by a different government in December 2014. The moratoria in place in Newfoundland and Labrador, and Quebec, along with the initial de facto moratoria in Nova Scotia, are similar to the state government of New York, which implemented a moratorium in 2011 (Rinfret, Cook, and Pautz 2014).

The variation in the Canadian context raises empirical questions regarding regulatory and policy design, such as: *Why do some provinces regulate some elements and not others? Why do some provinces produce comprehensive regulation while others regulate single issues? Why have some provinces issued moratoria on hydraulic fracturing while others have not?* The following chapter explores how these differences in regulatory frameworks – both across and within regions – provide an opportunity to explore why and how policy makers interpret and respond to the economic and ecological uncertainties of emerging technologies.

Chapter Three

Analysing Uncertainty

Fracking is a policy area that can function as either a low salience energy policy contending with distributive considerations or a high salience environmental policy reflecting regulatory politics (Lowry and Joslyn 2014; Lowry 2008; Davis and Fisk 2014). As with other policy areas straddling the intersection of technology and society, such as the regulation of genetically modified organisms, biofuels, windfarms, and assisted reproductive technology, fracking presents a series of uncertainties to decision makers regarding the extent and probability of economic benefits, ecological harms, and political losses. Indeed emerging scholarship on hydraulic fracturing suggests that as a result of the volatility of economic estimates and epistemic uncertainty regarding environmental risks, objective measures may have more limited explanatory power than *perceptions* of the severity of environmental harms or probabilities of economic benefits (Davis and Fisk 2014; Fisk 2013; Richardson et al. 2013; Small et al. 2014; Neville et al. 2017; Boudet 2019). Identifying the conditions that influence how decision makers come to determine the certainty – or lack thereof – of environmental harms is thus key to understanding variation in policy design (Neville et al. 2017). When economic benefits are perceived to outweigh environmental harms, we can expect single-issue or comprehensive regulation. Conversely when environmental degradation is considered more important than economic outcomes, precautionary measures such as moratoria and bans may be adopted.

Research in comparative environmental politics suggests that structural conditions have substantial weight in determining policy makers' decisions to limit or contain environmental harms. In Canada, industry elites have had a significant role in the politics of resource extraction (Hessing, Howlett, and Summerville 2005; Doern 2005; Carter, Fraser, and Zalik 2017; Urquhart 2018). Since hydraulic fracturing lies at the intersection of resource development and environmental policy, we would expect the preferences of industry actors, together with the ideological beliefs of a given party, to strongly determine the policy choices of government decision makers (Lindblom 1980; Carter 2020). Initial case studies of Saskatchewan and British Columbia suggest that economic and political dependence on the oil and gas sector explains minimal regulation of hydraulic fracturing (Carter and Eaton 2016; Carroll, Stephenson, and Shaw 2012). Scholars have also identified the relative strength of industry and partisanship as key factors influencing hydraulic fracturing regulation in US states (Weible and Heikkila 2016; Wiseman 2014). Fisk (2013) examines variation in the stringency of chemical disclosure regulation in thirty-four US states, finding partisan political leadership and relative income levels of the state influenced the stringency of regulation. In one of the more comprehensive quantitative examinations of heterogeneity in US regulation, Richardson et al. (2013) confirm that states with more oil and gas wells were more likely to regulate broadly across a number of elements, often with a focus on water quality and quantity. They found little support for the effect of political or environmental variables, such as the governor's partisan affiliation or existing environmental harms (e.g., number of species at risk) in their regression analysis. This finding is in keeping with Fisk's (2013) observation that objective levels of water pollution did not correlate with more stringent regulation in US states.

Together, these studies suggest several arguments. First, right-leaning provinces with an existing history of oil and gas regulation are more likely to regulate through piecemeal, single-issue regulations. Second, jurisdictions with a left-leaning governing party and a more limited history of development are more likely to pursue precautionary measures such as moratoria or bans.

Nevertheless, a simple analysis of interaction of industry and political ideology is unsatisfying in explaining the range of provincial regulatory variation in Canada. Recall that in 2012 Alberta introduced a more comprehensive framework than Saskatchewan, a time period in which both provinces were experiencing a significant boom in production and were led by right of centre political parties (Carter and Eaton 2016). On the other side of the country, in 2013 a Newfoundland conservative government introduced a moratorium on fracking, despite extensive experience with offshore oil and gas production, and a New Brunswick conservative government implemented a comprehensive set of rules. These empirical puzzles suggest a need to examine the limits of industry influence in shaping the preferences of policy elites (Baka et al. 2018).

Research on contentious politics in energy policy making suggests that when policy issues become salient they can capture elites' attention, making precautionary regulation possible (McAdam and Boudet 2012; Neville 2021). Although energy politics has traditionally functioned as a policy issue dominated by quiet politics (Culpepper 2011), scholars have documented a shift of energy policy making toward that of more salient environmental issues and increasing environmental activism (Boudet et al. 2014; Brown et al. 2014; Davis and Fisk 2014). In the United States, Davis (2012) finds that new coalitions of ranchers, county commissioners, hunters, and environmentalists, together with the partisan leadership of a Democratic governor helped explain Colorado's regulatory stringency in comparison to Texas. In a comparative case study of Colorado, New York, and Ohio, Rinfret, Cook, and Pautz (2014) also find that the influence of environmental and industry interest groups is mediated by the partisan preferences of state governors or legislatures. Other studies have also confirmed a rising politicization with regard to hydraulic fracturing in the United States (McGowan 2014), as well as attempts of environmental advocates to shift the debate to the national level (Davis and Hoffer 2012).

As environmental activism increases the profile of fracking in the public discourse, policy makers working on the file are more likely to be attentive to public opinion, creating opportunities for policy change (Baumgartner and Jones 1993; Soroka and Wlezien 2010). Similar to other environmental areas, research has found that public opinion regarding hydraulic fracturing is responsive to political cultural factors, with more highly educated, liberal women more likely to support moratoria than older white men (Andersson-Hudson et al. 2016; Brasier et al. 2013). Nevertheless, as with other contested energy issues (Gravelle and Lachapelle 2015), the direct impact of political ideology is mediated by other factors such as perceived distance/proximity and the distribution of material benefits (Clarke et al. 2016; Alcorn, Rupp, and Graham 2017; Kriesky et al. 2013; Boudet et al. 2018; 2016). Within this context, issue salience is likely to be influenced by the ability of environmental advocacy groups to open up public debate and generate contestation regarding potential environmental harms (Schattschneider 1960; Pralle 2006a).

Defining Risk Narratives

Environmental advocates can boost issue salience through the construction of policy problems, frames, stories, and narratives (Blyth 2001; Schmidt 2008; Béland and Cox 2010; Hay 2010; Béland 2010;

Hajer 1993). Problem definition is often a political struggle, with different sets of actors using discursive tools to construct and stabilize particular definitions of policy problems linked with preferred policy solutions (Cobb and Elder 1971; Stone 1989; Schon and Rein 1995; Stone 2011; Hoppe 2010; Gormley 1986; McBeth et al. 2007). Baumgartner and Jones (1993) put forward the concept of "policy images" to refer to the empirical and evaluative components that underpin actors' perceptions of a given policy issue (25). Policy images can include empirical observations about cause and effect, but they can also incorporate causal stories, narratives, and frames to elicit a normative tone (Baumgartner and Jones 1993). Policy images can maintain the status quo, reinforcing the stickiness of existing policy and regulatory frameworks. But when policy images are redefined in a way that increases issue salience and attracts the attention of a new set of actors, such as alternative experts, bureaucrats from different agencies, or opposing politicians, radical policy change can occur (Hall 1993; Baumgartner and Jones 1993; Pralle 2003; 2006a).

Empirical work on issue definition in fracking policy debates has focused primarily on the trade-offs between economic benefits, energy security, and environmental harms. Bomberg (2013) examines the different frames used by different "mobilization networks," tracing the difference in frames of economic growth, energy security, reassurance regarding technology, environmental and health risks, technological insecurity, and skewed benefits. In a case study of the adoption of disclosure rules in Colorado, Heikkila et al. (2014) examine changes in framing strategies over time, finding that interest groups are more likely to be polarized on policy core beliefs (e.g., pro/against development) than on secondary policy beliefs (e.g., pro/against chemical disclosure). Baka et al. (2018) find that narrative storylines in US congressional hearings reinforce oil and gas policy monopolies at the state level. Other studies have examined the influence of discourse and threat frames in hydraulic fracturing debates in the United Kingdom, Colorado, and New Mexico (Cotton, Rattle, and Van Alstine 2014; Jaspal and Nerlich 2014; Bomberg 2013; 2017a; Kalaf-Hughes and Kear 2018.

Building on these findings, I argue that specific ideas about uncertainty – which I term *risk narratives*[1] – are particularly effective in shaping policy makers' attention to environmental harms. Risk narratives are collectively held ideas that specify (1) beliefs regarding the certainty of knowledge (unknown risks) and (2) beliefs regarding the nature of the potential harm (dread risks). Risk narratives foster expectations regarding who is best situated to provide policy solutions and the most appropriate venue for policy formulation to take place.

There is strong analytical grounding as to why risk narratives might have such a potent function in hydraulic fracturing policy making. At the individual level, risk perception studies have found that risk can have significant influence on the ways in which different actors come to understand their policy preferences (Slovic 1987; Breakwell 2007; Taylor-Gooby and Zinn 2006; Gastil et al. 2011; Douglas and Wildavsky 1983; Kahan et al. 2010; Kahan 2015; Kasperson et al. 1988).[2] Research in cognitive psychology has measured risk perception along two different axes: (1) unknown risks and (2) dread risks (Slovic 1987). When individuals perceive *unknown risks* as high for a given technology, the technology poses new risks that are unknown to science, often because effects are delayed over time or harms are not observable. If *dread risks* are high, a technology is perceived to have great potential for generating uncontrollable, catastrophic harms that have an unequitable impact on different communities and may also pose significant risks for future generations. Dread risks are often perceived to be involuntary, in that they are imposed upon individuals with very little control or choice. Research in cognitive psychology finds that as our perception of dread risk increases, so does our desire for regulation (Slovic 1987; 1993; Sunstein 2005; 2009).

Political science research on hydraulic fracturing has found tentative support for the relationship between risk perception and support for environmental protection (Fisk 2013, Falkner and Jaspers 2012; Boudet 2019). In an early hydraulic fracturing policy study, Davis and Fisk (2014) theorize that the relationship between political ideology and the public's policy preferences is likely mediated through risk perception. Similarly, in his study of chemical disclosure rules, Fisk (2013) hypothesizes that lack of consensus among scientific actors as well as general limited information among the public are responsible for generating ambiguity surrounding risk perception, resulting in less stringent regulation. Among public opinion scholars, a burgeoning research agenda has emerged identifying the influence of individuals' prior beliefs, particularly culturally held perceptions of environmental risks, on support or opposition for hydraulic fracturing[3] (Andersson-Hudson et al. 2016; Clarke et al. 2015; Evensen et al. 2014; Boudet et al. 2014; Bullock and Vedlitz 2017; Brasier et al. 2013; Neville et al. 2017; Olive and Delshad 2017; Jacquet 2012; Jacquet and Stedman 2013; Montpetit and Lachapelle 2017; Howell et al. 2017).

At the policy subsystem level the salience of unknown and dread risks resonates with findings from comparative public policy scholarship on the role of scientific uncertainty and ambiguity in providing opportunities for ideational contestation and debate among opposing

Table 3.1. Types of risk narratives

		Dread risk	
		LOW	HIGH
Unknown risk	HIGH	COMPLEX	UNCERTAIN
	LOW	LINEAR	CATASTROPHIC

interests during policy formulation (Cairney, Oliver, and Wellstead 2016; Wellstead, Cairney, and Oliver 2018). In his examination of economic policy making, Blyth (2006; 2009; 2013) argues that there is an ontological distinction between conditions of risk, in which people rely on statistical calculations of probability, and conditions of genuine uncertainty in which individuals use alternative social and institutional practices, such as professional associations, or rules of thumb to govern future action. Under conditions of uncertainty, normative battles of persuasion, argumentation, and storytelling among policy elites, interest groups, advocates, and activated publics matter in determining our collective policy actions.

Drawing on the distinction between unknown and dread risks (Slovic 1987; 1993; 2000), I put forward a four-part typology of risk narratives in hydraulic fracturing policy debates: linear, complex, uncertain, and catastrophic (see table 3.1).

Linear risk narratives are those in which people collectively perceive both dread and unknown risks of a technology to be low. Technologies with linear risk are those that we are accustomed to, such as telephone transmission lines.

Complex risk narratives are those in which dread risks are perceived to be relatively low, but concern about unknowns, often phrased as scientific uncertainty, is higher. In these situations we assume that although there is some uncertainty about predicting outcomes, the extent of harm generated by a particular technology is within the realm of a tolerable risk. Air pollution from coal-fired electricity generation is an example of complex risk. Pollution thresholds are considered to be complex but possible to calculate drawing on environmental science.

Uncertain risk narratives are those in which perceptions of dread and unknown risk are high, reflecting a state of radical uncertainty about a given technology. Gene splicing, uranium mining, and geoengineering are examples of technologies that people have categorized as uncertain risks.

Finally *catastrophic risk narratives* are those in which actors perceive a particular technology as highly likely to generate serious dread risks – significant, involuntary harms that can have catastrophic impacts. For example, extreme harms that are certain to result from nuclear contamination would fall in the realm of catastrophic risk.

Risk Narratives and Hydraulic Fracturing

In sum, the analytical framework for this book rests on the assumption that the political economy of each province, and particularly the strength of the oil and gas industry, helps determine policy makers' attention to economic benefits over ecological harms. Yet because of the high degree of uncertainty regarding the new technology and because the policy area is nascent, I argue that interest groups' regulatory preferences are undetermined, meaning that signals to policy makers – and most important the salience of environmental harms relative to economic benefits – are not always clear.

In this uncertain context, different sets of actors within hydraulic fracturing policy subsystems will put forward various frames to advance their agenda: shale gas creates jobs; fracking is safe; fracking contaminates ground water; science is uncertain; fracking is a threat to human health, and so on. Although individuals, or particular interest groups, may stress specific elements of risk and uncertainty, I argue that collectively at the meso-level of the policy subsystem, these framings coalesce into a collectively held, dominant risk narrative. By tracing clusters of these frames and assessing the intensity of unknown and dread risks, we can identify whether the dominant narrative is of linear, complex, uncertain, and/or catastrophic risk.[4]

Although risk narratives emerge from particular social, economic, and political contexts, often shaped by individual actors' intentions, the prime aim of this study is to examine the impact of these narratives on policy making. I argue that risk narratives guide governments' actions, shaping who policy makers turn to for information and where policy formulation occurs. By examining the social construction of risk in each jurisdiction, dominant actors, institutional conditions, and the mechanisms by which policy change occurs, we can better elucidate why policy elites select various regulatory mixes in different jurisdictions at different times. Figure 1.1 in chapter 1 (page 9) provides a schematic overview of the different clusters of factors that influence variation in regulatory frameworks for hydraulic fracturing.

Risk narratives include claims about the extent of potential environmental harm and the certainty with which we can assume harm will occur. Different risk narratives have different capacities to generate issue salience, triggering urgency within political systems and attracting the attention of decision makers in different ways. As such, risk narratives function as an intervening factor, with an independent causal weight in explaining regulatory reform. I argue that risk narratives influence three key dimensions of power in regulatory decision making: (1) *who* participates, (2) *where* regulatory formulation happens, and (3) *how* regulatory design occurs through different mechanisms of learning.

Who Participates? Where?

Policy scholars have long noted that issue definition battles can expand the scope of conflict, engaging a broad range of actors into policy decision making beyond the classic trifecta of bureaucrats, interest groups, and politicians (Schattschneider 1960; Birkland 1998; Kingdon 1995; Baumgartner and Jones 1993; Hall 1993). New issue definitions can also precipitate a shift of decision making authority from within bureaucracies to new institutional venues (Baumgartner and Jones 1993). For example, advocacy groups can use new problem definitions to justify delegating policy making to a separate level of government in federal systems or to supranational governance institutions such as the European Union or the United Nations. Actors can also shift decision making laterally to markets or the courts. In these cases policy change is generated by the ability of actors to "venue-shop" – to move policy debate to a new institutional arena which challenges the legitimacy of the existing institutional arrangements of the policy monopoly, opening up possibilities for new policy development (Pralle 2003; 2006a; Hoberg and Phillips 2011; Thorn 2018).

The applicability of venue shopping has been demonstrated repeatedly in the US context as environmental scholars have documented the rapid proliferation of environmental policy and regulatory politics in the 1970s and 1980s (Repetto 2006). These scholars have noted the success of institutional venue shifting in fostering increased environmental regulation. For example, venue shopping by advocates to global networks and to municipalities has precipitated significant changes in forestry and anti-pesticide policies respectively (Pralle 2006b; 2006c; 2006a). In the Canadian context, although opportunities to venue shift are theoretically available to policy advocates, scholars have noted that environmental groups have historically been less successful in using the courts as a venue than their US counterparts, in part because of the more limited scientific expertise of Canadian environmental actors, but also because of constitutional constraints (Boothe and Harrison 2009; Harrison and Hoberg 1994). With regard to hydraulic fracturing in the United States, scholars have documented the attempts of environmental activists to shift the debate to the national level to counter the strength of pro-development state governments (Davis and Hoffer 2012). Advocates have also attempted to move policy making to the local level (Warner and Shapiro 2013; Christopherson, Frickey, and Rightor 2013) to try to diminish pre-existing policy monopolies between state executive administrations and oil and gas industry representatives (Baka et al. 2018). These studies suggest that changes in institutional venues may facilitate the adoption of more stringent hydraulic fracturing regulations in some cases depending on the ability of environmental advocates to increase their legitimacy as contributors to regulatory development.

Energy scholars have also noted the propensity of different administrative agencies, such as stand-alone oil and gas commissions, to limit the extent of regulatory change (Cook 2014; Hoberg and Phillips 2011). When institutional structures are more closed, either because of delegated authority to independent regulators, or because of tight policy networks made up of bureaucrats and government officials, policy change is more likely to occur at the level of instruments and settings, without any major deviations from the status quo (Skogstad 2008; Culpepper 2011; Hall 1993). Pierson (1993) refers to this as "institutional insularity," arguing that technical learning is more likely when decision makers are insulated, and when "a small number of actors are involved" (617). When institutional structures are more open to a broader set of actors, because of stakeholder consultation processes, court challenges, or municipal debates for example, more radical policy change, such as instituting a moratorium on hydraulic fracturing, is possible (Hoberg 2013; Atkinson and Coleman 1992; Hall 1993). These studies suggest that institutional arrangements can influence the scope of regulatory change by determining who participates in policy formulation and design.

We can thus classify venues for energy regulatory formulation as either closed or open. In closed policy subsystems the prime players are industry representatives, technical experts, and regulators (Hoberg and Phillips 2011). An example of a closed network is regulatory development within independent energy regulators, such as the Alberta Energy Regulator. Or venues can be open, as in the case of joint task forces or external review panels which encourage participation of a broader cross-section of society members, including local residents, Indigenous groups, environmental advocates, and academics from a variety of disciplines (Pralle 2006a; Skogstad 2008). The Nova Scotia Independent Review of Hydraulic Fracturing was an example of a more open network. Governance structures that are open are likely to incorporate a wider range of actors and knowledges (Dunlop and Radaelli 2018b), drawing on the expertise of epistemic communities (Haas 1992; Dunlop 2017), journalists (Sabatier 1988), and members of the public (Millar, Davidson, and White 2020).

How Does Regulatory Formulation Occur?

This study argues that risk narratives trigger different processes of learning within policy subsystems, shaping *how* regulatory change occurs. At a basic level, learning[5] is a mechanism through which policy makers update their normative and cognitive perceptions and behaviour based on past experiences, evidence from other jurisdictions, and scientific expertise (May 1992; Hall 1993; Weible 2008; Heikkila and Gerlak 2013; Zito and Schout 2009; Dunlop and Radaelli 2017; Heclo 1974). Policy learning scholarship has identified four different types of learning: (1) technical, (2) epistemic, (3) social, and (4) political (Hall 1993; May 1992; Bennett and Howlett 1992; Grin and Loeber 2006; Dunlop and Radaelli 2013; Heikkila and Gerlak 2013; Moyson, Scholten, and Weible 2017).

TECHNICAL LEARNING

In his analysis of paradigm change in macroeconomic policy in Britain, Hall argues that "normal policymaking," which is usually contained within the state and among government officials and scientific experts, can be understood as bounded within first or second order change in which policy goals are stable but incremental changes to policy instruments and settings can occur (Hall 1993). May also identifies "instrumental learning" as the process by which government actors demonstrate an improved understanding of the efficacy of particular policy instruments or implementation based on experience or formal evaluation (May 1992). Similarly, advocacy coalition framework (ACF) scholars argue that "policy-oriented learning" can lead to a gradual change in actors' secondary or policy core beliefs, reflected in changes in government policies (Jenkins-Smith, Nohrstedt, et al. 2014; Weible 2008; Heikkila et al. 2014).

EPISTEMIC LEARNING

Through the presentation of new scientific evidence, over time, some actors may be persuaded to adopt new views, at least related to secondary beliefs regarding policy instruments and settings (Rietig 2018). The ACF suggests that decision making guided by values is quite durable and resistant to change (Grin and Loeber 2006; Jenkins-Smith, Silva, et al. 2014). Policy core beliefs can only be changed very slowly over time through an iterative process of learning, usually dependent upon the introduction of new information or evidence within the coalition. New scientific information that does not unduly challenge actors' deep core beliefs may still serve to shift actors' secondary beliefs about effective policy options, leading to a change in adopted policy instruments and settings (Rietig 2018). Analytical work by Dunlop and Radaelli (2013; 2016; 2018a; 2018b) terms this type of learning epistemic learning, in which policy elites consult with scientific experts to inform regulatory design (Dunlop and Radaelli 2018a; 2018b).

SOCIAL LEARNING

Processes of social learning reflect a demonstrated shift in system-wide attitudes regarding a particular policy problem, specifically as to the normative dimensions of policy goals. Hall (1993) argues that social learning is indicative of a third order policy change, in which the policy subsystem experiences a paradigm shift in collective under-standings of the nature of the policy problem, the suite of solutions available, and the appropriate actors to drive implementation. There is some tension in the literature as to whether this type of process is best termed "learning" at all or is better understood as a form of political contestation. Although May (1992) argues that social learn-ing reflects a process of transforming the deep core beliefs of policy elites, Hall (1993) contends that paradigm shifts are triggered by an accumulation of policy failures, resulting in widespread contestation of the dominant paradigm through authority contests. Subsequent scholarship on social learning has tended to focus on the macro level, often examining paradigm changes using a historical institutional approach rather than focusing on belief change at the individual level (Grin and Loeber 2006; Skogstad 2011a; Moyson, Scholten, and Weible 2017; Dunlop and Radaelli 2017). Dunlop and Radaelli (2013; 2018a; 2018b) put forward the concept of "reflexive learning" in which radical uncertainty and open dialogue among teachers and learners interact to generate learning among a broad cross-section of society with regard to both policy goals and the instruments to achieve them. Learning in this sense is free form, unconstrained by hierarchy, exper-tise, or preconceived interests, resulting in redefined social norms, identities, and paradigms (2018b, 260). Studies of social learning have tended to focus on the ways in which members of the public become involved in policy making, as this type of learning tends to include a broader swath of society beyond policy elites (Hall 1993; Moyson, Scholten, and Weible 2017; Dunlop and Radaelli 2017; Trein 2018; Blyth 2013; Millar, Davidson, and White 2020).

POLITICAL LEARNING

Political learning is defined as "learning about strategies for advocating policy ideas or drawing attention to policy problems. The foci are judgements about the political feasibility of policy proposals and understandings of the policy process within a given domain" (May 1992, 339). The core distinction to be made between policy and political learning is that while policy learning is rooted in cognitive and normative beliefs regarding the content of the policy, political learning "is concerned with lessons about maneuvering within and manipulation of policy processes in order to advance an idea or problem" (May 1992: 339). Analytically, political learning is less concerned about change to the core policy preferences of decision makers and more focused on actors' perceptions as to the most viable or feasible means of advancing their political goals. Advocacy coalition framework scholars have identified similar mechanisms at play, termed "intra-coalition learning" within coalitions centred on political strategies, or "secondary beliefs" (Jenkins-Smith, Nohrstedt, et al. 2014). ACF studies have confirmed the proposition that intra-coalition learning is more likely at high levels of inter-coalition conflict, as actors within a given advocacy coalition aim to gain the upper hand (Weible 2008).

In his study of policy learning in the European Union, Radaelli (2009) defines political learning as incorporating strategic, substantiating, and symbolic uses of knowledge (1148), arguing that under conditions of political learning "we can expect better regulation policy goals to be set in accordance with their electoral feasibility, with policy performance as secondary goal ...[or] campaigns for better regulations which are highly publicized but not implemented" (1152). Returning to Heclo's analysis, Trein conceptualizes political learning as a form of powering, in which "collective actors learn new strategies to achieve their political goals," which Trein sees as rooted in striving for organizational legitimacy (Trein 2015; 2018). Dunlop and Radaelli (2018b) term this type of learning "learning as a by-product of bargaining" in which "decision-makers learn about the composition of preferences on an issue, the salient outcomes around which parties can coalesce and about breaking points – the red lines, so to speak ... [and] about the cost of reaching agreements" (262). Political learning is about policy design, but rather than focusing on the effectiveness of a given regulation, elites learn about the probable feasibility of policy implementation and reform.

Putting It All Together: Risk Narratives, Actors, Venues, and Learning

This book argues that risk narratives reflect shared causal and normative claims about hydraulic fracturing. These collectively held beliefs empower specific actors in the policy process and privilege specific institutional venues as the site of policy formulation. Together, these conditions facilitate different modes of learning that inform regulatory design. In some cases, risk narratives dampen the attention of both the public and regulators to environmental harms, fostering processes of technical learning to create single-issue regulations. In others risk narratives intensify the salience of dread risks, providing an opportunity for advocacy groups and members of the public to push for precautionary measures such as bans, generating a process of political learning among decision makers. The framework sets out four different clusters of narrative, actor, venue, and processes of learning with regulatory outcomes below (see figure 1.1, page 9).

Linear risk narratives reinforce the notion that risks can be easily managed within relatively closed venues, namely existing government processes and line departments. Linear risk narratives affirm the authority of regulators or technocrats to determine the appropriate course of action. In policy areas where linear risk narratives dominate, the prevalent norms, practices, and attitudes within the primary government agency managing the policy area likely determine the regulatory path (Renn 2008). Institutional characteristics such as the degree of discretion available to environmental or energy regulators (Bherer, Dufour, and Rothmayr Allison 2013; Huber and Shipan 2002) or the position of technical risk assessment within the existing governance hierarchy (Vogel 2012) are likely influential in determining regulatory responses to linear risks. When these narratives dominate, exploration of either harm or probabilities is thus likely to be contained among government actors in a relatively closed policy subsystem. Linear risk narratives can function as "cognitive locks," generating an institutional path dependency that reinforces particular causal ideas, generating self-reinforcing policy feedbacks and policy stability (Blyth 2001; Baumgartner, Jones, and Mortensen 2014; Jordan and Matt 2014). By fostering *technical learning*, linear risk narratives contain the scope of inquiry, leading to policy status quo or changes to instruments and settings, such as developing single-issue, stand-alone regulations.

Complex risk narratives legitimize policy makers to turn to scientific expertise to develop regulatory designs. Complex narratives reinforce the perception that state-academic collaboration is the best path toward mitigating harms. Policy studies have long highlighted the role of epistemic communities, especially academic scientific policy networks in providing "epistemic policy coordination" to states seeking to reduce uncertainty (Haas 1992; Dunlop 2017). Scholars examining a variety of environmental, health, and agricultural sectors have noted that both US and Canadian decision makers have tended to rely on scientific peer review as the most efficient and effective means of assessing risk (Hartley and Skogstad 2005; Jasanoff 2003, 229; Skogstad 2011b, 900; Vogel 2012). Blyth (2001) suggests that one of the mechanisms by which ideas determine institutional change is by their function as "blueprints" for the design of new institutions. Blueprints, Blyth stresses, help agents to redefine and better understand their interests under conditions of uncertainty. Blueprints are causal ideas that generate new instruments, such as ideas about dose-effect ratios and probability distributions of risk. Complex narratives thus legitimize certain sets of experts, such as hydrologists, geologists, or economists, and establish a privileged role for experts in the regulatory process. When complex risk narratives proliferate in a given policy area, decision makers are more likely to embark on epistemic learning driven by scientific experts in an attempt to reduce uncertainty and increase analytical tractability of a problem (Boswell 2009; Haas 1992; Neville and Weinthal 2016; Dunlop 2017). Complex risk narratives encourage *epistemic learning* between experts and government officials in a process that can broaden the scope of inquiry to encompass either single-issue or comprehensive regulatory frameworks (Renn 2008; Weible 2008).

Uncertain risk narratives in which both dread and unknown risks are high present decision makers with the challenge of determining the appropriate balance of precaution (Klinke and Renn 2012, 287). In the face of uncertain risks, scientific knowledge about the specific phenomenon under study is necessary, but also tenuous (Neville and Weinthal 2016; Dunlop 2017). Conditions of ongoing uncertainty can serve to undermine the credibility and validity of scientific and expert knowledge in the eyes of decision makers and the general public, limiting the ability of experts to provide new blueprints for action (Cann and Raymond 2018). Uncertain narratives justify an opening up of institutional governance, creating opportunities for social learning through public consultations and external reviews (Millar, Davidson, and White 2020). Open decision-making processes can increase public participation, including a broader range of groups, such as environmental and social advocates, representatives of Indigenous nations, as well as individuals, such as journalists and lay members of the public. Uncertain risk narratives have the potential to foster two different modes of learning. On the one hand, uncertain risks may trigger processes of *social learning* to mitigate deep epistemic uncertainty, a process which can broaden the menu of available policy options to include comprehensive regulation or bans. On the other hand, perceptions of ongoing uncertainty may prompt processes of *political learning* among elites to contain conflict, resulting in moratoria. In either case, regulatory development has the potential to incorporate new policy directions or goals (Neville and Weinthal 2016; Skogstad 2003; Hartley and Skogstad 2005).

Finally *catastrophic risk narratives* legitimize governments to act immediately and urgently protect the public interest. Under catastrophic risk narratives the driving actors in policy making are likely to be organized environmental interest groups who are able to mobilize the threat of broader public action, as well as elected government officials incentivized to maintain power (Dunlop 2017). Blyth (2001) stresses that once a degree of ambiguity is established in a policy area, different actors can use ideas as weapons to cast "previous institutional solutions as problems that [can] be diagnosed and cured only by their new ideas" (23). Ideas as weapons help actors outside of government, including interest groups, journalists, and experts, to mobilize action around a particular collective understanding of the benefits of a new institution or policy change (Baumgartner and Mahoney 2008; Mehta 2010; Béland 2005). Actors that are able to promulgate catastrophic risk narratives demand the attention of the electorate, sending a strong signal to policy decision makers to take rapid and immediate action (Slovic 1987). The venue for this type of regulatory development is likely to be in the public eye, but more adversarial as in town halls or public demonstrations. Although catastrophic risk narratives can theoretically create opportunities for political elites to engage in processes of *social learning* that foster more radical policy change, catastrophic narratives can also elicit processes of *political learning*, in which incumbent policy decision makers learn from other jurisdictions about the means of containing political conflict, or simply choose to act rather than be caught in a moment of indecision (Lesch 2018), resulting in the implementation of a ban.

Study Design

To identify the scope conditions and causal mechanisms through which regulatory change occurs, the study design focuses on regulatory change in four provinces: British Columbia, Alberta, New Brunswick, and Nova Scotia.

Case Selection

Following common practice for "small n" studies in comparative public policy, the study uses a "most similar" research design (Hall 2003), focused on a ten-year span from 2006 to 2016 to capture major regulatory developments over time.[6] Each selected case includes at least one instance of policy change. In British Columbia the regulatory framework shifted from existing rules for conventional oil and gas to single-issue regulation of hydraulic fracturing in 2012. In Alberta, the regulator adapted its conventional regulatory system to introduce a more comprehensive framework, also in 2012. In New Brunswick the framework transitioned from conventional rules for oil and gas to more comprehensive regulation of hydraulic fracturing in 2013. While in Nova Scotia in 2012 regulations changed from the existing conventional oil and gas framework to a moratorium. New Brunswick and Nova Scotia also included additional instances of change in regulatory frameworks across time, increasing variation. Although the study identifies broad differences between western and Maritime provincial policy, the analysis also focuses on regional regulatory variation over time, drawing comparisons between British Columbia and Alberta on the one hand and New Brunswick and Nova Scotia on the other. This regional comparison provides an opportunity to examine the specific conditions for detailed variation in regulation, as opposed to broader east/west variation in overall policy direction.

I have excluded specific analysis of Saskatchewan and Newfoundland and Labrador in the study, although I do consider the portability of the framework to these cases and the province of Quebec in chapter 8. As outlined in chapter 2 and identified by initial scholarship, Saskatchewan functions primarily as a negative case, having engaged in very minimal regulatory action on hydraulic fracturing, despite experiencing a significant boom in tight oil production throughout the 2000s (Carter and Eaton 2016; Olive and Valentine 2018). In their overview of regulatory variation in Canada, Carter and Eaton (2016) come to a similar conclusion, noting that Saskatchewan "opted to regulate fracking using existing regulations with very limited regulatory revisions in light of specific risks posed by fracking" (408). Since the study design is intended to explore regional variation, in this context the Saskatchewan case is likely to replicate dynamics in British Columbia. Similarly the moratoria announced in Newfoundland in 2013 is comparable to the Nova Scotia case, which serves to expand the case studies without adding significant additional variation to the study design.[7] In contrast, initial scoping for the project suggested that the Quebec case, although providing a first-mover example of government moratoria in Canada, has substantial differences in institutional regulatory structure from the rest of Canada, making it difficult to disentangle institutional and ideational factors. Scholarship on hydraulic fracturing in Quebec has noted the substantial influence of the Bureau d'audiences publiques sur l'environnement (BAPE)[8] in driving policy change (Dufour, Bherer, and Rothmayr 2011; Cleland et al. 2016; Chailleux 2020). Because the BAPE does not have a comparable provincial structure in the rest of Canada, a cross-provincial comparison is likely to identify governance as the prime mechanism of policy change, constraining the ability of the study to focus on the key factors of interest, namely the clustering of risk narratives, the dominance of the oil and gas industry, and environmental mobilization. The study excludes Ontario, Manitoba, and Prince Edward Island because of limited resource and policy development to date (Energy and Mines Minister's Conference 2013a). I use cross-case comparisons to confirm the presence and/or absence of factors conducive to policy change, such as the relative importance of the oil and gas industry in the political economy of the province, the degree of environmental mobilization, and the dominant risk narrative at play in each case.

Process Tracing of Causal Mechanisms of Policy Change

This book contributes to policy studies scholarship that documents and theorizes the complexity of policy making in which a multitude of actors are engaged in a variety of policy contexts (Cairney 2013; John 2003; Nowlin 2011). Following calls in policy studies to further refine the relationship between policy context, actors' choices, and mechanisms of policy change (Cairney and Weible 2017; Kay and Baker 2015), I examine the interaction of ideas, interests, and institutions in triggering policy change and identify different causal mechanisms through which policy change can occur.

This study defines a causal mechanism as a theoretical concept that refers to the pathway or process linking causes with particular effects or outcomes (Beach 2016; Gerring 2008; Checkel 2006). From the ontological perspective that causal mechanisms are more than mere intervening variables, the study also assumes that mechanisms are a causal chain or sequence that occurs through time and in which sequencing has a specific causal effect (Paquet and Broschek 2017). Within-case process tracing is used to identify "causal mechanisms that link causes (X) with their effects (i.e., outcomes) (Y)" (Beach 2016). Causal mechanisms here are the link between the "causes" of variation in industry strength, environmental mobilization, and risk narratives with the "effect" of variation in regulatory outcomes. I use *theory-building process tracing* as defined by Beach and Pedersen (2013) to gather evidence as to the presence (or absence) of four types of causal mechanisms: technical, epistemic, social, and political learning. In line with best practices in studying learning in policy processes (Heikkila and Gerlak 2013), interviews, direct observations, document and media analysis are used to identify the ways in which processes of learning influence individual and collective beliefs and behaviour, resulting in regulatory change (Heikkila and Gerlak 2013, 502).

Following Beach (2016) *pattern, sequence,* and *account* evidence support process tracing. Pattern evidence refers to the identification of patterns in the empirical record, such as frequency of occurrence of different frames in media debate for example. Sequence evidence refers to "making predictions about temporal or spatial chronologies of certain events" (Beach 2016: 469), for example tracing change in hydraulic fracturing regulation over time through document analysis. In contrast, Beach distinguishes account evidence as stemming from the specific content – or accounts – generated by actors or texts (Beach 2016). I draw on policy document analysis and media analysis to establish pattern and sequence evidence, and key informant interviews and media analysis to provide account evidence.

To gather evidence regarding regulatory change I analysed policy documents; Hansard; industry and environmental organization reports; scientific reports; think tank reports; and secondary literature in political science, environmental studies, and geography. I established a timeline of policy change in each of the cases and identified key moments of policy or regulatory change, examining patterns within and across cases. I used secondary and primary sources to identify the relative strength of industry and environmental mobilization in each province. I also analysed documents for support of the presence or absence of mechanisms of learning.

To gather pattern evidence as to the presence and type of risk narratives, I used in-depth thematic media analysis. For each province, I selected the local paper with the highest circulation and used the Canadian Newstand's database to search for relevant articles using the terms "hydraulic fracturing", "fracking", "unconventional gas", or "shale gas," from 2006 to 2016. The search yielded 218 articles from the *Vancouver Sun,* 246 articles from the *Calgary Herald,* 529 articles from the *Moncton Times and Transcript* in New Brunswick, and 264 articles from the *Halifax Chronicle Herald* in Nova Scotia. Sources were imported into NVIVO, a qualitative analysis software package and labelled according to year.

Following the research methodology developed by Lodge and Matus (2014), research assistants and I coded articles by highlighting claims in the text of the articles. Direct quotes or paraphrased comments that were directly attributed to actors by the article author were coded to eighteen different actor types (e.g., elected official, environmental advocate, industry representative, resident, or bureaucrat). Arguments or claims that were not connected to a specific actor in the article were attributed to the category "journalist."[9]

Highlighted arguments were then attributed to different environmental harms identified from the literature scan, such as "ground water risks" and "seismic risks." Arguments regarding potential benefits were also identified, such as "economic benefits," "energy security," and "clean energy." I also developed thematic categories inductively through the process of analysis that were grouped in broad baskets: (1) claims about unknown risks, including scientific uncertainty, and (2) claims about dread risks, including public concern. Each textual snippet, referred to as a "coding reference" in the analysis, was attributed to at least two categories: an actor type and a claim type. Some coding references were attributed to multiple thematic categories (e.g., elected official (opposition), unknown risks, and groundwater risks). Using a qualitative software package, I generated tables cross-referencing types of actors, claims, and dates. This analysis provides evidence as to the observable patterns in the types of arguments made by different actors at different times in the media during the period of interest.

To gather account evidence, I conducted thirty key informant interviews from 2014 to 2018. The study developed an interview schedule designed to probe the plausibility of the hypotheses regarding industry strength, environmental mobilization, risk narratives, and mechanisms of learning (see appendix 2). The study used snowball selection methods, together with searches of policy documents and government websites to identity key policy actors in each province. Interview questions were provided ahead of time to each interviewee, and interviews were conducted in person and by telephone using a semi-structured format, ranging from thirty to ninety minutes. Most interviews were conducted individually, I did conduct four group interviews with representatives from the Canadian Association of Petroleum Producers, the BC Oil and Gas Commission, New Brunswick Department of Health, and the Alberta Energy Regulator. In all four cases I spoke with bureaucrats involved in policy creation and environmental advocates providing policy advice. The interviews also included elected officials in New Brunswick, a political adviser in Nova Scotia, and industry representatives in Alberta and British Columbia. Interviews were recorded and relevant sections were transcribed.

Despite the efficacy of these methods, this study has some methodological limitations. First, it uses media debate as a proxy for public debate, supplemented by account evidence from key informant interviews. As policy scholars have noted, because of journalistic norms and practices, media debate can reflect a greater degree of polarization than that reflected in public opinion surveys or interviews with policy elites (Montpetit 2016; Crow and Lawlor 2016). From this perspective, the study is limited by lack of actual public opinion data testing for the effect of risk narratives specifically, although the research does include polling data from secondary sources. In addition, media concentration can diminish differences among provinces, making cross-case comparisons difficult. The study has tried to control for this by focusing on provincial papers rather than national outlets.

Second, account evidence from interviews is also limited in that interviewees were asked about risk narratives and learning mechanisms post-hoc, which can lead to incomplete data because of problems of recollection and/or bias (Moyson 2018). Information regarding the internal decision-making processes of elites is also limited because of a lack of access to elected officials in three of the cases (Nova Scotia, Alberta, and British Columbia). The study addresses this gap by examining public statements in the Hansard, personal blogs, and interviews with political observers to provide evidence as to revealed preferences of the actors.

Together, these methods elucidate the provincial politics of risk. Collective risk narratives emerge in different political contexts; these risk narratives guide provincial learning processes, ultimately resulting in different regulatory outcomes for hydraulic fracturing in each province.

Chapter Four

Limiting Uncertainty in British Columbia

In early 2011, Rich Coleman, the BC Minister of Energy Mines and Natural Resources and member of the incumbent BC Liberal Party, reassured British Columbians that they had nothing to fear regarding hydraulic fracturing.[1] In response to moratorium announcements by governments in Quebec and New York, he stated:

> I'm actually pretty comfortable with the maturity we have in this particular field ... I have seen nothing to date that would tell me that we are not out front on all the environmental issues compared to other jurisdictions. (J. Lee 2011)

Coleman went on record to assure residents that British Columbia had the "world's most stringent environmental regulations" (J. Lee 2011). The minister's confidence in the BC regulatory system illustrates the policy approach evident among other early adopters of hydraulic fracturing, such as Texas, in which hydraulic fracturing was considered to be an off-shoot of conventional oil and gas development. Coleman's statements reflect a narrative of linear risk; namely that the environmental, economic, and social risks of hydraulic fracturing were low and could be easily managed by technical expertise within government. The persistence of a linear risk narrative in the BC context is somewhat surprising for two reasons. First, despite Coleman's discussion of "maturity," exploration into British Columbia's gas reserves was relatively new in 2011 and had been in development for less than a decade. Second, by 2012 the salience of environmental risks from hydraulic fracturing in other jurisdictions had reached an all-time high.[2] Why did the narrative of linear risk dominate in British Columbia and what were the impacts on regulatory development in the province?

BC Regulatory Timeline

In 2002, following its 2001 election, the BC Liberal government introduced Energy for Our Future: A Plan for BC. The policy included incentives for the oil and gas industry, including specific support for coalbed
methane and unconventional resource development; streamlining of
environmental assessment, water licensing, and waste permitting; and
infrastructure development (Government of British Columbia 2002). In
2003 the government followed up with discounted royalties to encourage off-season drilling (S. Simpson 2006), and by 2005 several oil and
gas majors, including Royal Dutch Shell, had invested significantly in
unconventional gas exploration. In the summer of 2005, twenty rigs
were in operation. By the end of the year, the re-elected BC Liberals
reported an intake of $534 million in natural gas royalties for the year
(S. Simpson 2006; Whiteley 2005).

Throughout the latter half of the 2000s the BC government continued
to create incentives for unconventional gas development in the northeastern part of the province. In 2007 the provincial government released
The BC Energy Plan: A Vision for Clean Energy Leadership. The new
policy included a revised royalty framework with a variety of discounts
for unconventional gas production (Government of British Columbia
2007b; 2007a). In 2008 the province experienced a land boom from
investors seeking exploration licences in the Horn River and Montney
plays, an investment peak that observers attributed to the revised royalty regime as well as high gas prices (Palmer 2008). By December 2008,
the BC government had collected approximately $2.7 billion in royal-
ties and had begun to position itself as a potential exporter of liquefied
natural gas (LNG) to global markets (S. Simpson 2008). Following the
BC Liberals' third electoral victory in May 2009 and in the context of
falling gas prices on the North American market, in August the government introduced yet another revised royalty regime to stimulate further investment in northeastern British Columbia (S. Simpson 2009b).[3]
Prices remained low for 2009 and throughout 2010, resulting in a slowing of investment, although the province continued to report ongoing
investment in the Horn River and Montney plays (S. Simpson 2010).

In the context of falling gas prices, the BC government continued to
adapt its regulatory framework to incentivize both conventional and
unconventional gas development. In October 2010, revised regulations
under the Oil and Gas Activities Act (OGAA) came into force, further
streamlining environmental management, land, and water regulation
(Oil and Gas Activities Act, SBC 2008). The OGAA expanded the regulatory powers of the BC Oil and Gas Commission and included rules

pertaining to the following instruments: shale gas well data reports, minimum depth requirements for hydraulic fracturing, and the reporting of produced waters (Braul 2011). In the fall of 2011 Premier Christy Clark announced the government's intention to fully support LNG production, setting the goal of three operating LNG plants by 2020 (*Vancouver Sun* 2011).

On 3 February 2012, the government released two linked strategies under the general aegis of the 2011 BC Jobs Plan: The Natural Gas Strategy and the Liquefied Natural Gas Strategy (Government of British Columbia 2012a; 2012b). The Natural Gas Strategy positioned industry as a key driver of job growth and focused on maintenance of the province's existing royalty regimes, infrastructure programs, and streamlined regulation as mechanisms for encouraging sector growth, along with domestic and global market diversification. The LNG Strategy committed to the development of three LNG plants to be in operation by 2020 with the aim of accessing Asian gas markets. The strategy argued that BC's LNG exports would result in a global reduction of GHGs by replacing coal-fired plants. The strategy also reversed commitments to the self-sufficiency policy regarding hydro-electric energy introduced in the 2007 Energy Plan in order to service the growing LNG industry (Government of British Columbia 2012b). In 2013 the government established a new Ministry of Natural Gas Development (and Responsible for Housing), and in 2014 adapted tax rates for LNG to further encourage development (Graham, Daub, and Carroll 2017).

In 2012 the government also began to make some changes to the regulatory structure regarding environmental risks. In January the BC Oil and Gas Commission introduced a new mandatory requirement for producers to file information regarding the chemical composition of hydraulic fracturing fluids on fracfocus.ca, a new government website based on the US template of fracfocus.org developed by the Interstate Oil and Gas Compact Commission and the Ground Water Protection Council (BC Oil and Gas Commission 2011). In August the BC Oil and Gas Commission also amended its drilling and production regulation to introduce new well spacing requirements increasing the allowable number of wells within a target area (BC Oil and Gas Commission 2012c). During this time the commission was also identifying linkages between hydraulic fracturing and increased seismic activity in the Horn River Basin and released a report documenting the correlation in September (BC Oil and Gas Commission 2012a). Internally in 2012 the commission developed baseline studies of water usage and in December launched the North East Water Portal (NEWT), an online tool designed to help model water supply and demand for the entire region (BC Oil and Gas Commission 2012b). In April 2013 the commission released a report outlining its new "Area-Based-Analysis" approach to development approvals, which was intended to facilitate regional long-term development that also accounts for environmental and social outcomes (BC Oil and Gas Commission 2013). In April 2014 the BC Ministry of Health released a report on the second phase of its human health risk assessment of the impact of potential oil and gas activity in northeastern BC (Koope and Intrinsik 2015).[4] In May 2014 the government passed the Water Sustainability Act (WSA), modernizing the previous Water Act. One of the key provisions of the WSA was to create a legislative framework for managing groundwater in British Columbia, which had not been previously regulated (M.-L. Moore et al. 2015).

In 2015 the BC Oil and Gas Commission engaged the international firm Ernst and Young (EY) to review the current regulatory framework based on an assessment of three major risks: water use and protection; induced seismicity; and quality of life disturbances (EY 2015b). Overall, the review found that the scope of the existing regulatory framework was "robust" but made recommendations regarding changes to data collection and monitoring practices and expanding the settings of existing regulatory instruments (EY 2015b, 3).

The above timeline illustrates the ways in which the British Columbia government has pursued an incremental approach to policy and regulatory development vis à vis hydraulic fracturing. As identified in chapter 2, the government has largely pursued the gradual development of single-issue regulation (e.g., well spacing, chemical disclosure, groundwater monitoring), complemented by broad pro-development policy directives.

The Political Economy of British Columbia

BC's pro-development regulatory regime is perhaps unsurprising given the political economy of the province. Natural gas production was established in the northeast of the province in 1954 and throughout the second half of the twentieth century British Columbia was Canada's second largest producer of natural gas, contributing approximately 10 to 11 per cent of Canada's total gas production annually (CAPP 2014; Bott 2004). However by mid-2000s, producers and governments were anticipating an ongoing decline in conventional natural resources in western Canada, prompting increased attention to shale gas as a potential offset (NEB 2009). Exploration expanded rapidly in the space of a few years; by 2005 several oil and gas majors, including Royal Dutch Shell, had invested significantly in unconventional gas exploration in northeastern British Columbia. In the summer of 2005 20 rigs were in operation (S. Simpson 2006; Whiteley 2005); by 2011 the BC Ministry of Energy and Mines estimated that there were approximately 855 gas wells operating in the Horn River Basin, Montney play, Liard Basin, and Cordova Embayment (C. Adams 2011). In 2015 500 wells were drilled in the Montney play, reflecting 94 per cent of the total number of wells for the year (BC Oil and Gas Commission 2015). In 2012 production from shale and tight gas plays accounted for 50 per cent of total production (Stephenson, Doukas, and Shaw 2012), rising to 84 per cent in 2015 (BC Oil and Gas Commission 2015). Royalties on exploration licences boomed in the late 2000s, with a record high of $2.66 billion paid in land sale bonuses to the BC government in 2008 (C. Adams 2011). Production continued to increase to a high of 50 billion cubic metres in 2016, although royalties and land tenure leases have declined substantially due to low market prices for gas since the late 2000s (M. Lee 2017).

Beyond the salience of economic benefits from royalties, hydraulic fracturing production also presented significant potential to support a new LNG export industry in British Columbia. In the LNG Strategy (Government of British Columbia 2012b) the government estimated that the proposed three LNG terminals would generate "over $20 billion in direct new investment, as many as 9,000 new construction jobs, [and] about 800 long-term jobs" (4).[5] This rapid development, driven by demonstrated interest from foreign investors and major oil and gas companies, shifted the attention of the provincial government toward developing a pro-gas policy regime.[6]

Notwithstanding the strength of industry, the provincial government's unequivocal support of LNG development is somewhat puzzling, given the importance of environmental issues among the public in British Columbia. In general, British Columbian electorates tend to demonstrate a higher degree of concern for environmental issues than other western provinces. Polling conducted by EKOS Research (2016) finds that support for protecting the environment in the face of increased energy costs is 13 percentage points higher in British Columbia than in Alberta or Saskatchewan. British Columbia also has a strong environmental non-governmental organization (ENGO) network that developed during the anti-logging movement in the 1990s and that has been successful in mobilizing against pipeline developments in British Columbia and oil sands development more generally (Pralle 2006a; Hoberg 2013; Hoberg and Phillips 2011; Janzwood 2020).

A key area of uncertainty for the BC government has been an ongoing disconnect between the government's commitment to LNG development and its climate reduction goals (Stephenson, Doukas, and Shaw

2012; Stephenson and Shaw 2013; Carroll, Stephenson, and Shaw 2012). British Columbia has been a leader in developing subnational climate change policy, implementing a carbon tax in 2009 (Harrison 2012) and the *Climate Leadership Plan* in 2016 (Government of British Columbia 2016). The 2016 plan aimed to mitigate the impacts of shale gas development through incentives for reducing methane emissions and financial support for electrification of production and processing (Government of British Columbia 2016). However models developed by the Pembina Institute found that implementing two of the proposed LNG export facilities made it likely that British Columbia would miss its 2020, 2030, and 2050 methane emissions reduction targets (Heerema and Kniewasser 2017). It seems reasonable to assume that the contradiction between the government's economic goals regarding LNG and environmental goals for climate reduction, together with existing environmental mobilization networks, would have the potential to ignite public opposition, but the issue of hydraulic fracturing remains off the radar for most British Columbians. Despite the 2017 change in government from Christy Clark's Liberals to the NDP-Green coalition led by John Horgan, the commitment of the government to LNG development remains strong, with Premier Horgan asserting more recently that LNG development and climate action are not mutually exclusive (Cattaneo 2018). Why was it that fracking failed to generate the type of contentious politics in British Columbia evident in other Canadian jurisdictions?

Risk Narratives in British Columbia Policy Debates

Despite the general attention of the electorate to environmental issues, the dominant risk narratives in British Columbia did not coalesce around uncertain or catastrophic environmental risks. As a result, BC policy makers were able to discount potential environmental harms, focusing instead on economic benefits, buoyed by narratives of linear risk put forward by industry representatives. Together these storylines reinforced the position of a pro-development coalition in the province, limiting opportunities for environmental advocates to raise their concerns. Figure 4.1 presents an overview of the distribution of different frames across time in the *Vancouver Sun*; figure 4.2 reports different frames by actor types.

In British Columbia the pro-development coalition primarily comprised oil and gas industry representatives, together with a number of elected officials and some journalists. Similar to the political dynamics evident in New Brunswick (see chapter 6) and other natural gas producing states in the United States (Baka et al. 2018), the majority of pro-development arguments focused on economic benefits. In the BC case, much of the reporting focused less on potential future revenues and more on reporting of in-hand royalties and land tenure leases, especially in the late 2000s (Palmer 2011). For example, in March 2009, Minister of Energy, Mines and Petroleum Resources Blair Lekstrom commented:

> It's about more than dollars and cents, but if we are talking about the revenue side of it, this is by far the leading economic engine in the province and I think we should be thankful we have it ... the money is nice but the jobs it creates for the families in this province, in the northeast particularly, allow them to make a living and provide for their families and make their mortgage payments. So it's a positive all around ... I think we've reached a new level ... Our royalty credit programs ... put us at the forefront of the competitive nature of this industry. We don't compete with just Alberta or Saskatchewan. We compete with Texas. We compete with anywhere around the globe. (S. Simpson 2009a)

Leckstrom's enthusiasm for economic development would be later echoed by both Premier Christy Clark and Energy Minister Rich Coleman both of whom extolled the "exciting opportunities" of LNG development (Parfitt 2012), particularly in exporting LNG to Asian markets. The potential for economic gains was asserted by a wealth of oil and gas companies aiming to develop LNG export terminals in Kitimat, BC, including oil and gas majors such as Shell, Encana, and Petro-China (Vancouver Sun 2012). The assertion that British Columbia was a major player in natural gas production would be later complemented by pro-LNG narratives that highlighted the potential for LNG exports to reduce greenhouse gas emissions by facilitating other countries' transitions to low carbon energy sources (Coleman 2013; Chen 2020a; Stephenson and Shaw 2013). The Canadian Association of Petroleum Producers, an umbrella association for the oil and gas industry, also linked frames regarding clean energy with energy security, reinforcing the argument that unconventional gas production would lead to North American self-sufficiency (Vancouver Sun 2013).

Beyond promoting economic benefits and energy security narratives, industry representatives were quick to assert the safety of hydraulic fracturing as an industrial process. The Canadian Association of Petroleum producers, for example, repeatedly raised a talking point that "170,000 wells have been fracked in Alberta without any adverse effects" (Morton and Healing 2012). The assertion that hydraulic fracturing was a safe, proven technology that had been used for decades

Figure 4.1. Risk narratives by year in BC debates

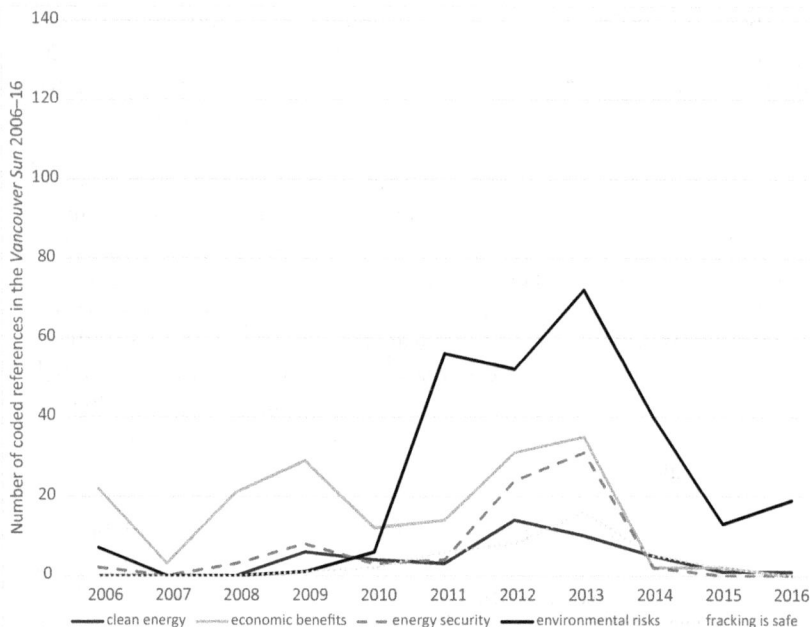

reflects a narrative of linear risk, which in turn justifies an absence of regulatory reform. For example, Doug Bloom, then president of Spectra Energy Transmission, a natural gas pipeline company, noted, "We know it has become an issue elsewhere, but frankly, fracking is not a new technology ... We've been fracturing wells in Western Canada for decades and to my knowledge there haven't been any problems associated with that" (J. Lee 2011). Industry claims about the safety of fracking implied that unknown risks were low because fracking had been in operation for decades and dread risks were low because there had not been any ecological impacts. Figure 4.2 illustrates the similarity between the claims put forward by industry representatives and elected government officials, with both sets of actors drawing on arguments regarding economic benefits and the linear risk of the technology.

In contrast to the multi-faceted nature of the narratives put forward by government officials and industry representatives, the anti-fracking framing that emerged in the province focused predominantly on environmental risks. As figure 4.1 illustrates, arguments regarding

Figure 4.2. Risk narratives by actor type in BC debates

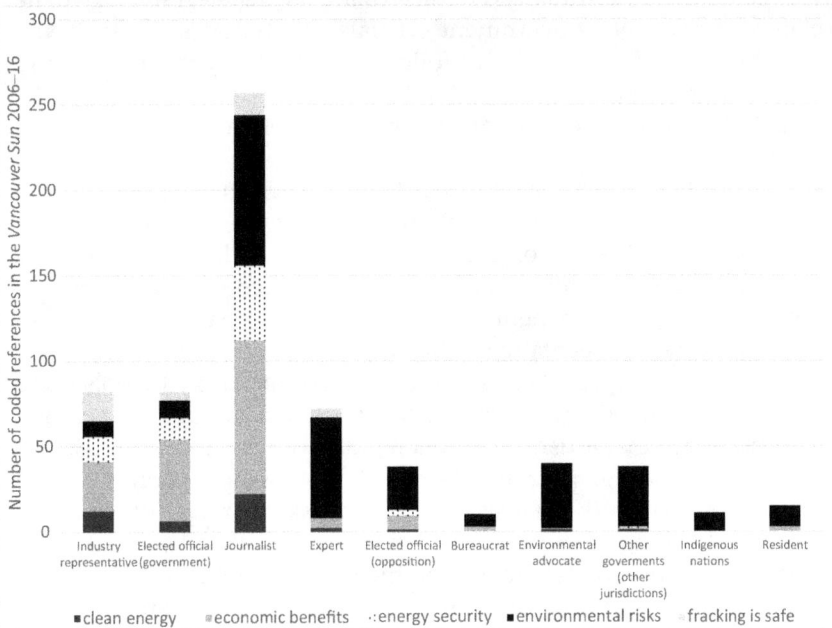

environmental harms emerged in 2010 and peaked in 2013 during the British Columbia provincial election. As in other jurisdictions, concerns regarding water – particularly groundwater contamination – were more salient than other risks. However, potential risks to climate change also began to emerge in 2012–13, particularly with regard to the impact of the LNG export terminals on British Columbia's GHG emission reduction targets (Pynn 2013). Apart from discussion of the emissions from LNG export terminals, most arguments regarding environmental risks were tied to experiences in other jurisdictions, as opposed to specifically referencing BC habitats (see figure 4.2).

Anti-fracking frames regarding environmental risks were most likely to be put forward by environmental advocates, members of Indigenous nations, and other residents, as well as elected officials from the BC NDP (in opposition) and the BC Green Party. Reports from academics were also more likely to highlight environmental risks than did government officials or industry representatives. Overall, the media analysis of the *Vancouver Sun* illustrates the relatively low salience of the issue among policy actors in British Columbia. Unlike coverage in the Maritime provinces, most British Columbia coverage was generated by journalists characterizing the debate, rather than the claims of residents, government officials or advocates. As such, the distribution of frames tends to follow journalistic norms of objectivity by reiterating both pro-development and anti-fracking arguments, resulting in a relatively equal distribution of claims supporting and opposing development.

In sum, the BC media analysis demonstrates the prevalence of a dominant narrative of linear risk, namely that the likelihood of ecological harms in the BC context was low or that, at the very least, there were favourable environmental trade offs between potential harms to groundwater and the benefits of clean energy. This narrative was complemented by arguments regarding various economic benefits, either through direct investment and job creation or indirectly by bolstering North American energy security. The narrative of linear risk included assertions that the British Columbia regulatory system was robust and could easily manage potential harms. The following sections examine the ways in which this narrative of linear risk prompted processes of technical learning among bureaucrats in the BC Oil and Gas Commission and dampened the attention of the general public to the issue. As such, the BC case illustrates that linear risk narratives can align well with closed policy subsystems in which the bureaucracy is a dominant actor.

Linear Risk and Technical Learning

Unlike jurisdictions in eastern Canada, where responsibility for regulating oil and gas development tends to be split among departments of natural resources, energy, or the environment, regulatory authority in British Columbia has been delegated to an independent oil and gas regulator. The BC government established the BC Oil and Gas Commission in 1998 as a crown corporation with the aim of being a "one-stop shop" to streamline environmental assessment, licensing, and other regulatory requirements for producers (BC Oil and Gas Commission 2014a; Parfitt 2011; Graham, Daub, and Carroll 2017). Significantly, at the same time that hydraulic fracturing was emerging as a regulatory issue in the late 2000s, the government was also finalizing legislation (*Oil and Gas Activities Act, SBC* 2008) designed to further delegate authority to the BC Oil and Gas Commission, enabling it to amend and create new regulation. According to Paul Jeakins, former commissioner and CEO of the BC Oil and Gas Commission, the revisions to OGAA provided the commission with the flexibility to consider substantial regulatory changes without engaging cabinet or the provincial legislature:

> It was absolutely serendipitous that that [the revision to the OGAA] was already underway. When you look at that, 2007 was when we started to get a real inkling that this was going to change conventional [gas production] to something called unconventional and the Act was brought down in 2008 – it was well underway, that took years to do, and all the regulations had the benefit of looking at the shale development. (Jeakins 2015)

The institutional independence generated by the changes to the OGAA provided actors within the commission with the regulatory space to consider adapting regulation on their own terms, in support of, but separate from broad policy directions developed at the Ministry level. The increased delegation of authority to the commission insulated bureaucrats from political interference (Jeakins 2015; confidential interviewee 2015c) while also providing the opportunity to be more responsive to the demands of industry (confidential interviewee 2015c).

The institutional insularity of the BC Oil and Gas Commission was complemented by framing strategies pursued by elected government officials throughout the early stages of shale gas development in the province to distinguish between LNG and hydraulic fracturing, despite the obvious practical connection between the two. As the media analysis above demonstrates, the main arguments in favour of development centred on domestic job creation and international reduction of GHG emissions through trade with Asia. Ministers and elected officials were also much more likely to use the term "LNG" to refer to overall policy goals of clean energy and investment, reserving "hydraulic fracturing" for discussion of specific regulation, such as mandatory chemical disclosure. Stephenson et al. (2012) note that clean energy and the notion of gas as a transitional fuel to reduce climate change emissions is a predominant frame in the 2012 Natural Gas Strategy. My own textual analysis of the 2012 Natural Gas Strategy and the 2012 Liquefied Natural Gas Strategy finds that LNG is mentioned 104 times, while "hydraulic fracturing" is mentioned only 14 times.

The distinction between the two terms had a twofold interpretive effect. First, the positive framing of LNG bolstered the legitimacy of government policy at the Ministry level, generating ongoing support for the policy even in the face of catastrophic risk narratives emerging from US media debates (Stephenson, Doukas, and Shaw 2012; Stephenson and Shaw 2013). Second, containment of discussion of hydraulic fracturing to technical complexities at the regulatory level served to further dampen public attention to the issue by reinforcing the narrative of linear risk. The narrative that the harms (if any) of hydraulic fracturing could be managed deftly and competently by adequate regulation, enabled regulators to work slowly on incremental changes to regulation without political intervention. The ideological distinction between LNG policy and fracking regulations as separate issues reinforced the closed policy subsystem of industry representatives and experts within the BC Oil and Gas Commission. These actors were thus tasked with dealing with the (perceived to be) predominantly technical issue of extracting the resource responsibly from the ground (confidential interviewee 2015c).

Because of the distance between LNG policy developed at the Ministry level and the regulatory framework managed by the BC Oil and Gas Commission, political pressures on the regulator during this time were limited (Jeakins 2015). As a result, technical experts and policy analysts within the commission developed regulatory responses at an incremental pace, beginning with well spacing in 2008 and moving on to seismicity and chemical disclosure by 2012 (confidential interviewee 2015c; BC Oil and Gas Commission 2012c; 2011). These changes to regulation occurred over a five-year period, with analysts drawing on experiences of regulators in the United States through British Columbia's position as an international member of the Interstate Oil and Gas Compact Commission (IOGCC)[7] (confidential interviewee 2015c; Jeakins 2015). The most obvious case of policy transfer from other jurisdictions is the commission's adaptation of fracfocus.org to manage chemical disclosure. Fracfocus is an online database initially developed by the IOGCC in conjunction with the Ground Water Protection Council designed to provide public access to chemical additives used in fracturing on a well-by-well basis in the United States (Dundon, Abkowitz, and Camp 2015). In 2011 the commission began working with the IOGCC to develop a database for the Canadian context, which was launched in January 2012 as fracfocus.ca (Government of British Columbia 2012c). As Jeakins noted:

> An early innovation was fracfocus. That was a part of the water discussion, this disclosure of hydraulic fracturing fluids is getting out of control, make it mandatory and put it online. And the conversation just went ... We were working with the Interstate Oil and Gas Compact Commission that started it, so we just went down there (we are associate members) and we started working closely with them, used their website, and brought it up here, made it available to the rest of Canada, saying we're doing this, we're going to make it mandatory, over to you. If you want to be on it, great ... it helps because we have that ability to create our own regulation

for I don't want to say the technical stuff, but it is technical regulations, the pipeline regulations, and I think that model works because it makes us much more nimble versus going through the whole Cabinet process, so our experts can react very quickly. (2015)

The commission also strengthened in-house technical expertise through two major research projects: examining instances of induced seismicity in the Horn River and Montney Shale and ground water mapping through the development of the North East Water Portal. Working with data from Natural Resources Canada and in consultation with geoscientists from the University of British Columbia and the Alberta Geological Survey, the commission released studies on the Horn River Basin and Montney shale in 2012 and 2014 respectively (BC Oil and Gas Commission 2012a; 2014c). Although these studies confirmed the link between fluid injection and seismic activity, similar to reports in Oklahoma (Andrews and Holland 2015), as figure 4.1 demonstrates, the empirical data did not prompt the same degree of media attention as the Oklahoma study would in 2016. Instead, the BC case reflects the characteristics of quiet politics (Culpepper 2011): The structure of the commission, facilitated by the 2008 amendments to the OGAA, created the space for bureaucrats to engage in technical learning based on policy evidence from epistemic communities and other jurisdictions and generated by their own researchers in concert with industry.

Absence of Salient Catastrophic Risks

One of the empirical puzzles of the BC case is why hydraulic fracturing has failed to become a prominent concern among the general public. Despite existing environmental networks, anti-fracking mobilization has been much more limited in degree than in the Maritime provinces. From an ideational perspective, the strong delineation between LNG and hydraulic fracturing made it difficult for anti-fracking activists to link environmental risks with LNG development in the public consciousness. As one interviewee noted:

Whenever I would try to do media around LNG I would try to talk about fracking, and vice versa, [with fracking] I would talk about LNG, and I remember one instance when the journalist did not understand the connection. The Christy Clark government was pushing LNG, LNG as the economic saviour a boom for the province. So LNG was all over everything, and she never talks about fracking. We've been trying hard

to connect those dots, but probably the general public and even journalists – they probably understand a bit better now – but that connection
was not fully understood. (confidential interviewee 2018)

As noted above, the separation between the economic and climate benefits of LNG and the environmental risks of hydraulic fracturing has
remained surprisingly durable in the public consciousness in British
Columbia; in 2016 an Insights West poll found that while 43 per cent of
British Columbians supported LNG, only 23 per cent support fracking
(Insights West 2016).[8] The construction of LNG as having both economic
and climate benefits also served to legitimize the BC government's position as a climate leader while unequivocally pursuing shale gas production. In their analysis of the intersection of BC's carbon pricing initiatives
and oil and gas production, Houle, Lachapelle, and Purdon (2015) argue
that the rapid expansion of shale gas in the province helps explain the
retreat of the government from entering emissions trading markets with
California and Quebec under the Western Climate Initiative, which were
"perceived as being at odds with the economic interests" of the province
(51). As such, the BC case is somewhat of an early litmus test for the interpretive malleability of LNG, as a potential "clean" resource in contrast
to the "dirty" construction of "fracked gas," foreshadowing interpretive
politics that would later emerge regarding bridge fuels in Canada (Clarke
et al. 2015; Neville 2021; Janzwood and Millar 2022).
    In contrast to the position of elected officials, established
environmental organizations in British Columbia initially focused on
groundwater risks and climate impacts of hydraulic fracturing. In 2011
the Pembina Institute issued a series of reports evaluating potential
risks with regard to water and climate (K. Campbell and Horne 2011),
following up with a national discussion paper and policy leaders' forum
in 2012 (Pembina Institute 2012; Angen and Switzer 2012). The Pembina
Institute conducted further analysis of the impact of LNG development
on BC's capacity to meet its GHG reduction targets in 2014 (Goehner and
Horne 2014), building on earlier academic studies conducted in 2010
(Jaccard and Griffin 2010). In 2014 three well-established environmental
organizations – Ecojustice, the Western Canada Wilderness Committee,
and Sierra Club BC – launched a suit against the provincial government
for permitting short-term water withdrawals for hydraulic fracturing
(Canadian Press 2014). Although the case was dismissed on the grounds
that the volume of water used did not unduly diminish overall supply
(D. Moore 2014), it demonstrates environmental advocates' attempts
to increase the salience of the issue by shifting the debate to another
venue.

Interviewees noted that despite evidence that the regulatory framework was lacking with regard to groundwater protection, the discussion of risks to water failed to generate the groundswell of opposition it did in eastern provinces such as Quebec (confidential interviewee 2015d; 2018). Part of the challenge in mounting a campaign against groundwater contamination in British Columbia was the perceived distance – both physically and psychologically – of communities and environmental advocates from the source of industrial activity. Unlike the geography of Quebec, for example, where urban communities were much more likely to be situated close to a potential development (Montpetit, Lachapelle, and Harvey 2016), many urban dwellers in British Columbia lack awareness of the northeast of the province entirely. As one interviewee commented, "New Brunswick takes a few hours to drive across, but to get up to Fort St. John from Vancouver is two days of driving. So it's not in most people's imagination even" (confidential interviewee 2018). As a result, groundwater risks, while highly concerning in New Brunswick and Nova Scotia, failed to generate the same type of catastrophic risk narratives among the BC electorate at large, thereby limiting attention to directly affected communities in the northeast of the province and reducing the opportunities for mobilization. More recently, advocacy campaigns have focused more specifically on the land/health impacts from "liquefied fracked gas" terminal sites on the coast in an aim to find a framing of the policy problem that is more relevant to the public (Sierra Club BC 2018). As an interviewee noted, "to be honest, there have been intentional attempts to find things that will resonate more broadly. If climate's not the thing, and if fracking in the north east is not the thing, then let's talk about something that will freak out population centres" (confidential interviewee 2018). The lack of proximity of urban British Columbians to potential environmental harms served to reduce the immediacy of the perceived risks, reducing the salience of the issue.[9]

In the absence of a catastrophic risk narrative, anti-fracking activists were also hampered by a greater degree of fragmentation among different groups in comparison to mobilization in the Maritime provinces. Some of the most vocal advocates against both hydraulic fracturing and LNG development have been members of the Fort Nelson First Nation who have traditional lands in the northeast of British Columbia governed by Treaty 8 (Garvie and Shaw 2014). Members of the nation have raised significant concerns regarding groundwater contamination and cumulative land and health risks impacting their treaty rights (Garvie and Shaw 2016; Rayher and Gillis 2015). Nevertheless, as policy scholars Garvie and Shaw (2016) document in their case study of the Fort

Nelson First Nation and other Treaty 8 First Nations, many community members also held the perception that development is inevitable, stating "[there is a] common perception that if a community says 'yes' to some development so that community members can benefit economically, the environmental impacts are an inevitable repercussion that must be borne" (1018).

These dynamics have also played out among Indigenous nations throughout British Columbia. For example, another strong group advocating against LNG has been the hereditary members of the Gilseyhu organizing at the Unist'ot'en camp located on Wet'suwet'en territory near Smithers, however some of the elected officials of the Wet'suwet'en have provided consent for LNG pipelines to go through the territory (McSheffrey 2015). Similarly, the Nisga'a Lisims Government has also supported LNG pipeline development (Meissner 2014) despite opposing bitumen pipelines a few years earlier (Noble 2012,Terrace Standard 2012). In effect, the incentive for Indigenous nations in British Columbia to maintain openness to economic development opportunities in the future has led some to support LNG as the "lesser of two evils" in comparison to bitumen pipelines, which are perceived to have greater land and health impacts and fewer economic benefits (confidential interviewee 2018). The more limited anti-fracking unity among Indigenous nations is also reflected in the opposition of settler communities in the northeast. Unlike the Maritime provinces, and especially New Brunswick, where local rural residents quickly mobilized against hydraulic fracturing companies, in northeastern British Columbia there is substantial support for the oil and gas industry, as evident in the results of the 2013 election when the pro-development BC Liberals captured over 57 per cent of the popular vote (CBC News 2013a). This fragmentation was also evident among environmental advocates and local governments. In their survey of policy actors in the BC policy subsystem, Montpetit et al. (2016) found that approximately half of local governments and a quarter of environmental groups expressed support for shale gas development in the province.

The higher degree of fragmentation among anti-fracking advocates in British Columbia was compounded by a lack of support from the official opposition party, a factor which became crucial in the 2013 provincial election. In their 2013 platform the opposition the BC NDP came out strongly against the Enbridge pipeline, asserting that the party would "withdraw from the NGP [Northern Gateway Pipeline] equivalency agreement on environmental assessment with the federal government within one week of forming government" (British Columbia New Democratic Party 2013, 41). This position was strengthened

Figure 4.3. Dread and unknown risks by year in BC debates

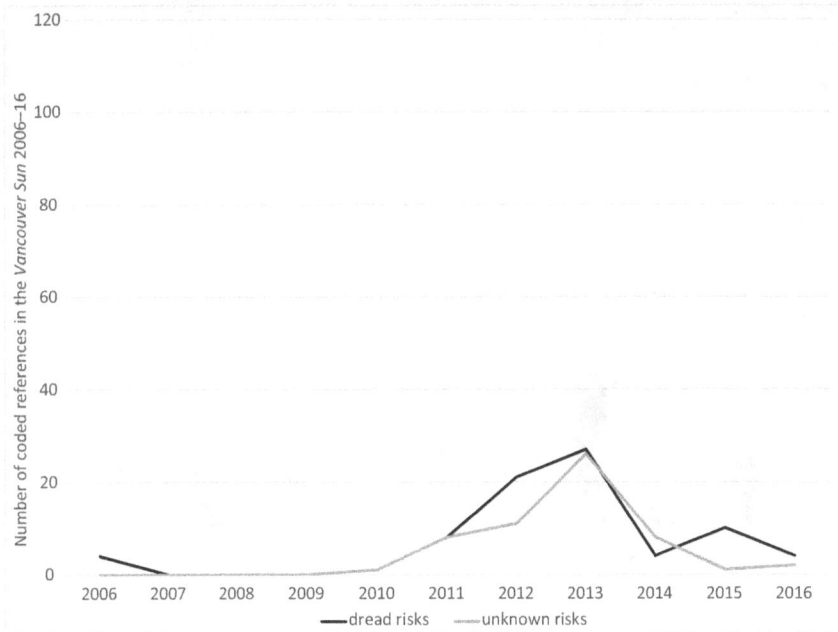

by statements from BC NDP leader Adrian Dix during the campaign; he expressed his opposition to the Kinder Morgan Pipeline expansion as well (Hoberg 2013). Election observers note that this shift ultimately put Dix on the defensive during the campaign because it positioned the party as against economic development (Hoberg 2013; confidential interviewee 2015d; 2018). The strength of the BC NDP's opposition to pipelines made it very difficult for Dix to also oppose LNG development outright, similar to the trade-off faced by Indigenous groups between opposing the oil sands and supporting LNG. As a result, the BC NDP landed on a somewhat ambiguous position of calling for an "independent, expert-led public review of fracking" but stopped short of calling for a moratorium (British Columbia New Democratic Party 2013, 41). Figures 4.3 and 4.4 demonstrate the frequency with which references to scientific uncertainty became a talking point for members of the BC NDP in distinction to statements made by bureaucrats, elected officials in government, and environmental advocates.

The emergence of a narrative of uncertain risk is very similar to the position put forward by New Brunswick Liberal challenger Brian

Figure 4.4. Dread and unknown risks by actor type in BC debates

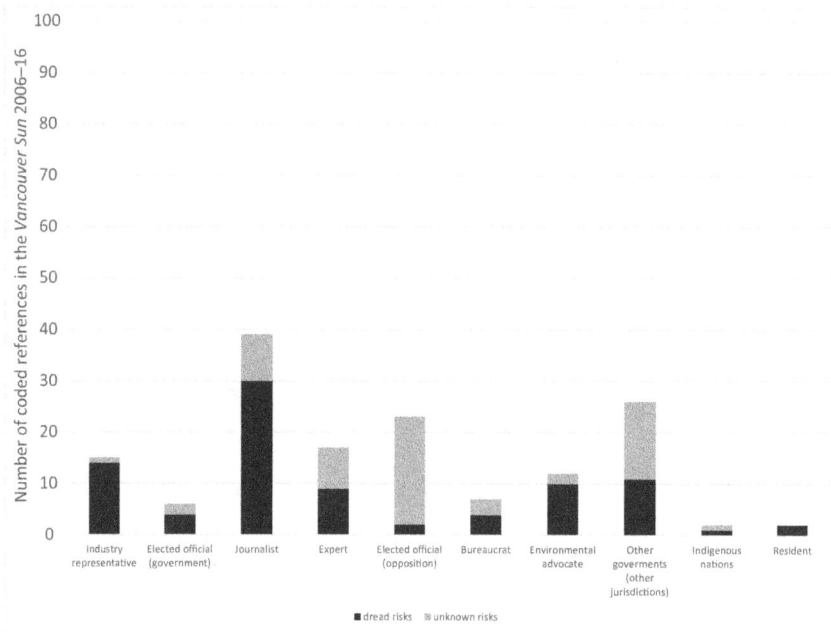

Gallant in the 2014 provincial election, but with a very significant difference. Although Gallant repeatedly used the context of scientific uncertainty to justify a moratorium (Government of New Brunswick 2013c), Dix went out of his way to emphasize that the party was not calling for a moratorium prior to the review, at one point demanding a retraction from a candidate who had suggested otherwise (Palmer 2013; Fowlie 2013). Ultimately the uncertain risk frame failed to mobilize public attention as it did in New Brunswick, either because of the dominance of pipeline debates in environmental issues, or because of the stability of the Liberal LNG narrative of economic benefits and minimal risks. As summarized by a *Vancouver Sun* editorial:

> The recent provincial election boiled down to a fight between jobs and the environment. The Liberals, with their focus on jobs, liquefied natural gas and exports to Asia, won. The NDP, with their emphasis on the environment and opposition – real or perceived – to pipelines, tanker traffic and natural gas fracking, lost. (Melton and Peters 2013)

Unlike the sequence of events in the Maritime provinces, the narrative of linear risk has proven to be durable in the BC context. In the absence of catastrophic risk narratives that linked causal concerns around groundwater contamination with substantial policy action such as moratoria or bans, authority contests in British Columbia have been subsumed to debates at the local level, mostly in the northeast. Without the mobilization of a multifaceted coalition drawing on a broad base of public opposition, hydraulic fracturing has failed to galvanize the attention of policy elites, limiting possibilities for political learning. Figure 4.4 demonstrates that the majority of references to dread risks, including public opposition and lack in trust in government, were actually put forward by journalists rather than Indigenous nations, other residents, environmental advocates, or elected officials in opposition. That these concerns were highlighted in the media and yet failed to generate salience among policy elites suggests the limits of the influence of media debate. As Montpetit, Lachapelle and Harvey (2016) argue, media actors can generate a burst of attention to prompt the disproportionate attention of elites; however, I would also add that this is only true if images of catastrophic risk generated by the media are taken up by activated publics.

The BC case demonstrates how linear risk narratives serve to facilitate technical learning but dampen social learning opportunities. As such, this case provides analytical nuance to the conditions laid out by Pierson (1993) for policy learning. The institutional autonomy of the BC Oil and Gas Commission has indeed been a crucial element in enabling the slow development of single-issue regulation, insulating the regulator from political debate. Yet although the BC case confirms the importance of institutional insularity for policy learning, the case also highlights the need to develop more nuanced definitions of policy complexity. Pierson argues "a second factor of significance [for policy learning] may be policy complexity. The greater the technical proficiency required to understand an issue and possible policy responses, the greater the likelihood that learning effects stemming from social investigation and analysis will be prominent" (618). Although Pierson defines policy complexity as including issues in which "causal chains are complex and uncertain," my research demonstrates the value in parsing out complexity from uncertainty. Although complexity may justify processes of policy learning, this effect is likely to hold only to a point – that in which the risks identified are presumed to be the purview of government scientists and technical experts, tolerably managed within government organizations. After a certain threshold of complexity, governments will be forced to cast the net wider to include other epistemic actors, widening the potential for contention and debate.

Chapter Five

Monitoring Uncertainty in Alberta

Alberta is the largest oil and gas producer in Canada and has significant experience in hydraulic fracturing in conventional oil wells, dating back to the 1950s (M.-L. Moore et al. 2015). The unprecedented boom in bitumen production in the province's oil sands in the early twenty-first century has been made primarily through in-situ extraction in which wells are "injected with steam or solvents" (Hoberg 2021, 18). To achieve these technological advances, the Albertan provincial government has developed a regulatory framework that privileges economic production, from investing for over thirty years in research and development, to creating favourable royalty regimes, streamlining regulatory approval processes, and reducing the scope of environmental oversight (Carter 2020; Urquhart 2018; Hoberg and Phillips 2011; Salomons and Hoberg 2014).

In an extensive overview of the interaction of Alberta's investment in the oil sands and its increasingly anaemic environmental protection regime from 2005 to 2015, Carter (2020) finds Alberta's regulatory framework to be significantly lacking in attention to cumulative effects, public participation, and ongoing environmental monitoring. These conditions would suggest that regulators and policy makers in Alberta were well positioned to simply echo BC's single-issue approach to hydraulic fracturing regulation. But a closer look finds that Alberta regulators chose to explore implementing a more comprehensive approach termed "play-based regulation (PBR)" (ERCB 2012a). The Alberta regulator also issued two new directives (59 and 83) focused specifically on chemical disclosure and well-to-well communication (ERCB 2012c; AER 2013a). Together, these regulatory components place Alberta closer to a more comprehensive regulatory framework, or at the very least, a "single-issue plus" framework in comparison to British Columbia (Carter and Eaton 2016). This chapter examines why the Albertan government and the regulator engaged in this significant overhaul of the regulatory system for unconventional gas and finds that catastrophic risk narratives prevalent in the United States, together with a contingent institutional restructuring, triggered a process of political learning among Albertan regulators, leading to a more comprehensive regulatory approach.

Alberta Regulatory Timeline

In January 2011, following an internal review of its regulatory framework for coalbed methane, shale gas, and tight gas, the Energy Resources Conservation Board (ERCB), the independent regulator for oil and gas in Alberta at the time, released an initial report detailing the characteristics of regulatory regimes in eight other jurisdictions in Canada and the United State (ERCB 2011).

Toward the end of 2011 Energy Minister Ted Morton announced that the government was initiating a "regulatory overhaul" focused on developing best practices in hydraulic fracturing for the ERCB. The government was concurrently considering the development of a "single-desk" regulator (Penty 2011b), and a year later Premier Redford introduced the Responsible Energy Development Act (REDA) for discussion and debate in the house. REDA aimed to create a "one-stop-shop" regulator for the oil and gas industry by combining the regulatory functions of the ERCB with relevant elements housed with Alberta Environment (Henton 2012; EY 2015a; Gerein 2012; AER 2013b). REDA was passed in November 2012 by the Progressive Conservative government and was proclaimed in June 2013, establishing the Alberta Energy Regulator (AER) and replacing the ERCB (Gerein 2012; AER 2013b). As with the restructuring of the Oil and Gas Activities Act in British Columbia, which enabled the commission to add new hydraulic regulations without returning to provincial cabinet, the process of restructuring provided the opportunity for the regulator to develop an incremental package of hydraulic fracturing reforms drawing on findings from other jurisdictions with little political interference.

In December 2012, the ERCB released Regulating Unconventional Oil and Gas in Alberta – a Discussion Paper outlining a new regulatory approach to address hydraulic fracturing that included "play-focused regulation" (ERCB 2012a), an approach which was subsequently termed "play-based regulation" (PBR). PBR was a regulatory regime that aimed to coordinate industry proponents on an area-based, rather than well-by-well approach. Key aspects of the regulatory design included closer attention to cumulative effects, increased expectations for stakeholder engagement, and streamlining of the application process (AER 2014;

EY 2015a; AER 2016). The ERCB invited written feedback on the new approach from citizens and stakeholders (ERCB 2012b). The AER subsequently piloted its new play-based regulation (PBR) in the Duvernay shale play near Fox Creek, Alberta (AER 2014; EY 2015a; AER 2016). In addition to play-based regulation, the regulator also issued two new directives focused specifically on chemical disclosure and well-to-well communication. Directive 059, Well Completion and Data Filing Requirements was issued in December 2012 and required producers to disclose chemical composition of fracture fluids and water use on fracfocus.ca (ERCB 2012c). In May 2013 the ERCB issued new regulatory requirements for managing subsurface well integrity under Directive 083, Hydraulic Fracturing – Subsurface Integrity (AER 2013a). During the same time period Alberta Energy Regulator also developed new processes for stakeholder engagement (Larkin et al. 2018; Millar 2021). When considered as a suite of changes, this regulatory reform reflects a more comprehensive response to hydraulic fracturing than British Columbia, a move which is puzzling considering the strength of the oil and gas industry in the province (Carter and Eaton 2016).

The Political Economy of Alberta

Alberta's position in the federation as a region rooted in both fossil fuel extraction and conservative politics is well documented by Canadian energy and environmental scholars (Macdonald 2020; Urquhart 2018). Alberta's oil sands position Canada as the holder of some of the largest oil reserves globally, with exponential growth since the early 2000s to 2.9 million barrels per day annually (Hoberg 2021). The structural power of the oil and gas industry in Canada has led scholars to characterize the Alberta as a subnational "petro-state," exhibiting classic characteristics of resource-dependent polities such as price volatility, social inequality, and the perception of oil production as a public good (Carter 2020, 13–14).

Yet despite, or perhaps because of extensive public and private investment in Alberta's oil sands, multi-stage horizontal hydraulic fracturing of shale gas was still in its infancy in the first decade of the millennium. A significant number of gas wells only began to be completed using this technology in the late 2000s, a few years after British Columbia (Lucas, Watson, and Eric 2014). Early estimates identified over ten shale plays in the province, concentrated in the Duvernay Region (Larkin et al. 2018; Rokosh et al. 2012). In 2013 analysts estimated that companies had invested approximately $1 billion in drilling activity, with 195 wells drilled (Healing 2013). Subsequent studies by the Alberta Geological

Survey estimate the Duvernay region has 76.6 Tcf in marketable gas (NEB and Alberta Geological Survey 2017).

With regard to electoral pressures, in Alberta, the public generally tends to demonstrate high levels of support for the oil and gas industry; a 2014 poll commissioned by the government reported 92 per cent of Albertans consider Canada to be lucky to have the oil sands (Innovative Research Group 2014). Alberta residents also tend to place a lower premium on environmental issues with just over half seeing the environment as a very important priority (Canadian Election Survey 2008). Despite this general support for industry, there has been localized environmental mobilization against hydraulic fracturing in the province, particularly against shallow fracturing in coalbed methane deposits. The most prominent voice has been that of Jessica Ernst, a former Encana employee who filed a series of legal challenges against Encana and the ERCB regarding groundwater pollution in Rosebud, Alberta (Nikiforuk 2015; Carter and Fusco 2017).

Risk Narratives in Alberta Policy Debates

As to be expected in Canada's premier "petro-province," the pro-development coalition in Alberta was driven primarily by the oil and gas industry, centred on a narrative of linear risk. Coverage in the *Calgary Herald* was attuned to the details of the shale gas boom not only in Canada, but in North America, Europe, and the United Kingdom. Early coverage focused on the potential of the US boom and potential economic impacts on Canadian industry, but often included greater detail as to industry specifics, with articles highlighting the pace of US development, reporting on the financial statements of various industry players, including both production and service companies, a distinction that was rarely reported on in the other provinces. As in British Columbia and New Brunswick, industry representatives were quoted often, although in Alberta a higher proportion of arguments stressed the safety of the practice of hydraulic fracturing, with oil and gas CEOs often reiterating the maturity of the Albertan oil and gas regulatory regime. For example, in response to concerns among Albertan ranchers in 2011, Mike Dawson, the President of the Canadian Society of Unconventional Resources commented: "[what] we need to recognize with oil and gas development, whether conventional or unconventional, is that the government has put in a strong set of regulations"(Penty 2011a). Industry lobby organizations such as the Canadian Association of Petroleum Producers consistently stressed that the practice of hydraulic fracturing was a "proven technology" with minimal risks to Albertans (Wood 2012).

Yet in comparison to British Columbia where officials often reiterated BC's "world class" practices, Albertan-elected officials were much less likely to be quoted in the media, leaving the regulator (first the Energy Resources Conservation Board, later the Alberta Energy Regulator) to comment. For example, in response to 2011 reports on well contamination in the United States, ERCB spokesperson Bob Curran reasserted a linear risk narrative, stating: "there has never been a documented case of fracturing activities contaminating groundwater in Alberta ... The ERCB protects groundwater with strict regulations to prevent gas migrating into groundwater sources" (Henton 2011). In comparison to the BC case, in which elected officials perpetuated a clean energy LNG "bridge" narrative early on to justify development, claims regarding both clean energy and energy security were less salient in Alberta and reflect a greater similarity to the Maritime cases than the province's geographic neighbour. When news coverage touched on clean energy, articles tended to report on the clean energy benefits of the shale boom in the United States, rather than in the Albertan context. The similarities in discourse between Canadian oil and gas leaders and US pro-development advocates illustrates the significant attention of Canadian oil and gas producers to the dimensions of public discord in the United States and the overall dominance of industry representatives in news media debates (see figure 5.2). The pro-hydraulic fracturing coalition in Alberta was thus somewhat similar in composition to British Columbia, with industry representatives and journalists asserting the economic benefits and relative environmental safety of the extractive practice and putting forward a narrative of linear risk. At the same time, the debate was much less politicized in Alberta, with fewer public statements from elected officials in government or in opposition.

Nevertheless, a surprising finding of the Alberta case is the higher overall salience of environmental risks in media dataset. The centrality of the oil and gas industry in Alberta meant that journalists were quick to cover environmental failures in the United States, specifically epistemic and popular reports of groundwater risks, well blowouts, inter-well communication, and seismic activity. The attention of journalists to US debates is evident in the prevalence of environmental risks in the dataset with their frequency in the *Calgary Herald* reaching similar levels in 2012 to the New Brunswick and Nova Scotia cases (see figure 5.1). As in the Maritimes, most of the claims in the dataset of environmental harms referred to groundwater or drinking water risks, however there was also significant coverage of seismic risks throughout the period of study. The *Calgary Herald* covered quakes in Ohio as well as an early report on seismicity by the US Environmental Protection Agency (EPA) in 2012,

Figure 5.1. Risk narratives by year in Alberta debates

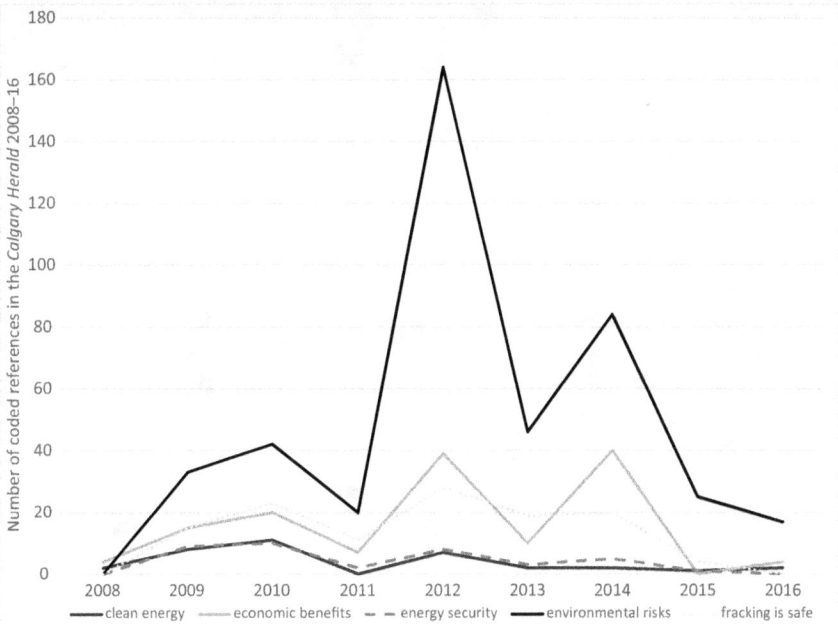

and later covered a series of quakes in 2014 in Oklahoma but also in Fox Creek, Alberta, where the AER was piloting PBR. The focus on different jurisdictions was more wide ranging in the *Calgary Herald* than in the *Vancouver Sun* across the period of study, including references to New York State, Pennsylvania, Colorado, Ohio, Wyoming, and the United Kingdom, as well as Quebec, Nova Scotia, New Brunswick, and British Columbia (see figure 5.2). The prominence of environmental harms in the media dataset suggests that catastrophic narratives were more present in Alberta than in British Columbia during the period of study.

However, in contrast to the Maritime provinces, the attention to environmental risks in the Albertan news media was driven primarily by journalists reporting on US public contention and environmental debate, rather than investigations of homegrown environmental mobilization against fracking. Although there was some coverage of localized opposition against coalbed methane production in Lochend and Jessica Ernst's case,[1] a strong Alberta-focused narrative of complex or catastrophic risk did not emerge among residents, environmental advocates, and Indigenous groups as it did in the Maritime provinces

Figure 5.2. Risk narratives by actor type in Alberta debates

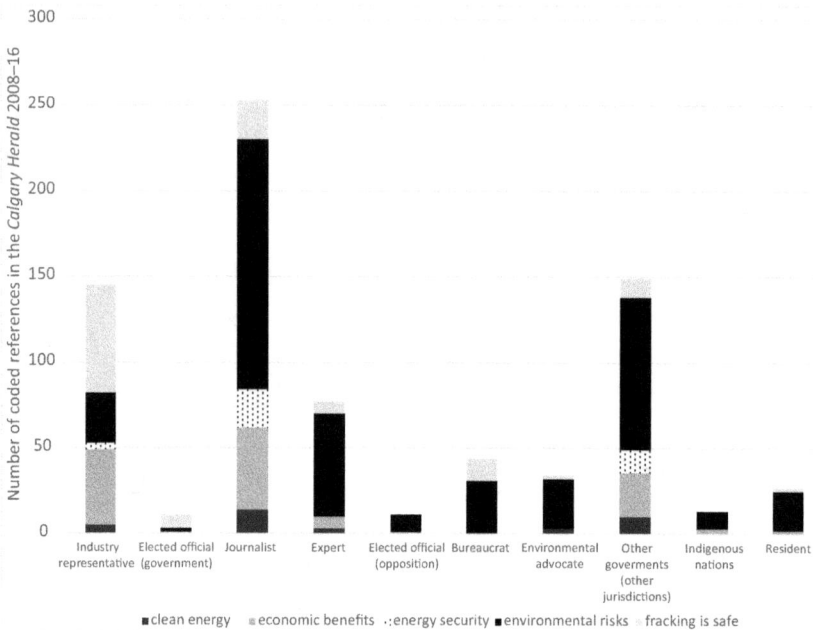

(confidential interviewee 2015a; Penty 2011b; Calgary Herald 2012). Instead, the attention to US environmental harms provided nuance to the linear risk narrative put forward by industry officials and regulators. Industry representatives were quick to assert the overall safety (and lack of novelty) of hydraulic fracturing, but both the Canadian Association of Petroleum Producers and the Canadian Society of Unconventional Resources were also highly aware of the need to address the public concerns regarding dread risks that had been steadily increasing in US news and social media throughout 2010 and 2011 (confidential interviewee 2015e). In early 2012, CSUR Vice President Kevin Heffernan commented:

> For the most part, the industry's practices and benefits were understood within the community it operated in ... But now it's much bigger and much more visible. People from every walk of life are making observations and drawing conclusions because there is an information gap. We weren't especially skilled at communicating what we did, and that is recognized by the industry. You will see over the next year a lot more public communication happening. (Hussain 2012)

In an interpretive dynamic that would be echoed by industry concessions to climate change in the second half of the decade, industry representatives were willing to acknowledge that in the face of public fears of dread risks, industry needed to develop better communication strategies, including packaging of regulations and public outreach activities. Thus, the risk narrative that came to dominate in Alberta can best be understood as mostly linear but with some contained or neutralized narratives of catastrophic risk. This duality of risk narratives would inform the processes of technical and political learning among policy makers, leading to more comprehensive regulatory reform.

Linear/Catastrophic Risk and Technical/Political Learning

The interpretive dynamics in Alberta initially set up a narrative of linear risk, which is to be expected in a petro-state. Government officials within the ERCB were keeping abreast of technical developments in the United States throughout the 2000s and by 2008 they were aware that horizontal drilling and hydraulic fracturing was quickly becoming the norm among operators in Alberta (confidential interviewee 2015b). In response, the ERCB undertook a three-pronged approach in 2010. Internal task groups would examine regulatory risks and emerging scientific developments while another team would review regulatory developments in other jurisdictions (ERCB 2011). Initial risk identification highlighted the policy complexity of the issue, both from a scientific perspective with regard to addressing subsurface integrity (including well-to-well communication) and groundwater monitoring, as well as from a regulatory management approach with regard to water withdrawals, well spacing, and production intensity (ERCB 2011). The findings of the internal review helped inform the development of the PBR and Directives 59 and 83 (confidential interviewee 2015b).

To formulate new standards, regulators at the ERCB relied on informal and formal connections with regulators across Canada and the United States. In Canada, top level officials in Alberta, Saskatchewan, and British Columbia participated in the Western Regulators Forum, an intergovernmental platform designed to harmonize oil and gas regulatory frameworks across the provinces (Government of Canada 2018). Alberta has also been an international member of the IOGCC since 1996 (IOGCC n.d.; 2018). The jurisdictional review team consulted with regulators from British Columbia and Saskatchewan, as well as Michigan, Louisiana, Pennsylvania, Oklahoma, Texas, and New York (ERCB 2011). Government officials within the ERCB drew in part on BC's experiences managing well-to-well communication as well as

Figure 5.3. Dread and unknown risks by year in Alberta debates

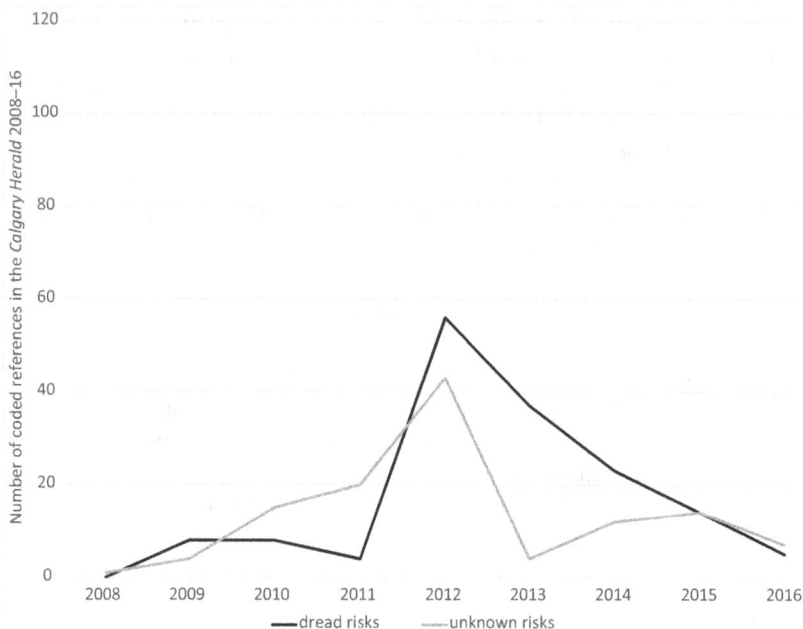

industry input in the development of Directive 83 (ERCB 2011; AER 2013a; confidential interviewee 2015b). Similarly, examination of BC's management of water withdrawals also suggested regional approaches to water management (ERCB 2011), an element of regulatory design that was ultimately incorporated into the cumulative effects objective of the PBR (ERCB 2012a; EY 2015a; AER 2016). For bureaucrats within the ERCB (and later the AER), hydraulic fracturing was simply another layer to the existing regulatory framework for the oil sands, which had already been developed based on unconventional technologies. From this perspective, developing new regulations was primarily a process of technical learning focused on tweaking individual elements to better reflect the practical challenges of hydraulic fracturing.

However, the Alberta case also illustrates the ways in attention to dread risk can spur processes of political learning among regulators, leading to more expansive regulatory change. Between 2008 and 2010, the staff at the ERCB were increasingly aware of the growing public concern and media attention to the potential for hydraulic fracturing to contaminate groundwater (confidential interviewee 2015b). In 2010, US documentary filmmaker Josh Fox released *Gasland,* a feature length film focusing on the contamination of wells in local communities in Pennsylvania and the growth of fracking in the US oil and gas industry (Fox 2010). The film generated considerable attention in both the US and Canadian contexts by providing an opportunity for activists to highlight environmental and health risks (Vasi et al. 2015; Eaton and Kinchy 2016). Albertan regulators also documented an increase in public inquiries to its regional offices at this time, with residents raising concerns regarding the higher degree of industrial intensity around shale gas wells, evident in increased trucking for example (confidential interviewee 2015b). This local attention was complemented by increased media attention to policy failure in US states, driving up issue salience with the Alberta regulator. As one government official noted:

> To be fair to Albertans too, there were a number of start-up problems in the United States. Several US states had no history of oil and gas, they had no foundation nor experience. No foundation of regulatory oversight, no effective service industries, et cetera. In that regard there were some documented problems that we would like to think would never happen here because of our rules, there's always compliance and assurance, but the rules are in place to avoid some of the situations that triggered the media coverage. But we had to explain that to Albertans. So part of this is the technical shift, some of it's the regulatory process shift, and some of it is a continued and accelerated engagement and involvement of the public. (confidential interviewee 2015b)

The ERCB jurisdictional review also highlighted public concern as an issue, wryly noting "Increasing public, media, and government attention is being focused on the potential for hydraulic fracturing of shale gas reservoirs to contaminate useable water aquifers with fracturing fluid chemicals and natural gas, despite no proven cases of hydraulic fracturing of deep zones having caused such a problem" (ERCB 2011, 5). The report subsequently refers to the New York moratorium and the use of chemical disclosure rules as key regulatory responses to addressing landowner concerns (ERCB 2011, 29). Regulators were also aware of the 2011 moratorium in Quebec (confidential interviewee 2015b). Together with stories of policy failure from US states, concerns from other Canadian jurisdictions served to drive up issue salience with the regulator.

In contrast to other provinces, industry players in Alberta were also very attuned to the potential for public mobilization to derail development, with industry acknowledging the need for improved

Figure 5.4. Dread and unknown risks by actor type in Alberta debates

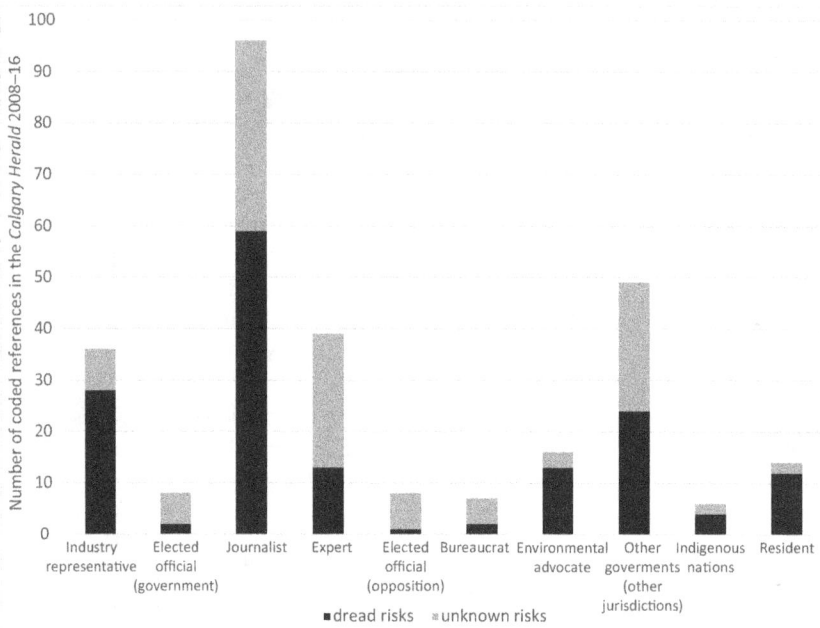

communication with the broader public around the practice (Yedlin 2015). Figure 5.4 illustrates that a high proportion of industry representatives were aware of the proliferation of "dread risk" among activated publics and were worried about rising public concern. Oil and gas firms in Alberta had already experienced a significant economic and reputational blow from environmental campaigning throughout the late 2000s that had successfully labelled bitumen production from Albertan oil sands as "dirty oil." The "tar sands campaign" had successfully linked anti-fossil fuel organizing across international borders around the dread risks of climate change. Ultimately, the tar sands campaign effectively lobbied the Obama Administration in the late 2000s to deny approval for the development of Keystone XL, a project designed to move Albertan bitumen to tidewater (Janzwood 2021; Hoberg 2013; Hoberg and Phillips 2011; Hoberg 2021; Janzwood 2020). As such, industry representatives were attuned to ways in which rising public concerns could force governments to attend to environmental harms. To ward off the danger of moratoria or bans, industry representatives consistently signaled their support for regulators to engage with the public to dispel organized opposition. The bureaucrats at the ERCB were also well aware of this need. As one respondent described:

> Compared to other jurisdictions in Canada, we have an advantage here because of decades of experience. Alberta has been looked to as leading regulation – there's certainly a culture of that – but there's an expectation that the regulators will stay on top of technological changes and emerging issues in communities and keep the regulatory framework up to date. (confidential interviewee 2015b)

The threat of burgeoning negative public opinion throughout the United States and Maritime provinces provided regulators with insight into *what not to do*: maintaining the status quo had left US and Canadian regulators susceptible to contentious politics. Within this context, the ERCB's decision to join British Columbia in using fracfocus.ca and the introduction of PBR provide an example of political learning, or in this case, negative lesson drawing. To avoid public outcry regarding water contamination snowballing in other jurisdictions, regulators ratcheted up the stringency of regulation, akin to the process that Hoberg (1991) refers to as activist-driven learning.

The Alberta case illustrates the ways in which catastrophic risks can influence regulatory design, even in Canada's petro-provinces. Concurrent with the development of the PBR and Directives 59 and 83 the government was also engaged in substantial institutional reform and redesign as a result of REDA. The establishment of the new Alberta Energy Regulator provided an institutional opening for the government to signal an increased attention to stakeholder engagement, a shift that was exemplified in the creation of a new Stakeholder and Government Engagement Division within the organizational structure (AER 2018). In effect, the institutional reorganization provided a window of opportunity for government officials to rebrand and emphasize that the regulator was responsive and attentive to public concerns (confidential interviewee 2015e; 2015a). As Stone (2017) observes, "negative lessons can have symbolic value and power in de-railing the proposals of opponents" (61). Environmental scholars have emphasized that REDA both streamlined the approvals process for industry and eviscerated the ability of third parties to obtain standing in regulator hearings (Ecojustice 2013; Carter, Fraser, and Zalik 2017; Carter 2020); at the same time, the AER piloted the PBR in Fox Creek in 2014 (ERCB 2012a; AER 2016).

But although the attention of media, industry representatives, and regulators to public concerns created a potential political opportunity for homegrown environmental mobilization to build, a province-wide anti-fracking movement did not emerge in Alberta during this time. Environmental networks in the province had been supportive of early organizing against fracking for coalbed methane in the early 2000s, identifying potential risks to groundwater and highlighting the need for water monitoring (Pembina Institute 2007). But as in British Columbia, by 2010 the resources of many environmental organizations in Alberta, including both established and grassroots movements (Janzwood 2021) were devoted to tar sands campaigning, making fracking much less salient in the Albertan context (confidential interviewee 2015a).

The Albertan case presents a compelling example of the deftness of advanced petro-states to navigate the interpretive politics of oil and gas development. On the one hand, the risk politics in Alberta were very similar to British Columbia – that hydraulic fracturing was simply another unconventional method of development, providing solid economic benefits to the already flourishing economy firmly rooted in oil sands development. On the other hand, the maturity of the industry also meant that both firms and regulators were much more attentive to the potential for successful challenges from environmental movements than their BC counterparts. While BC elected officials aimed to ward-off public concern through assertions of "world-class regulations," the regulators in Alberta aimed to reduce the scope of contention through institutional change, increasing pathways for stakeholder engagement through REDA and piloting play-based management. Because of previous high-profile environmental conflicts in the public domain, regulators were attuned to the ways in which narratives of catastrophic risk could mobilize public support and block development. In a relatively muted, yet steadily intensifying critical atmosphere, regulators implemented a regulatory regime designed to demobilize public opposition. As such, the Albertan case supports a link between catastrophic narratives and political learning, a dynamic that would also be crucial in determining regulatory outcomes in the Maritimes a few years later.

Chapter Six

Managing Uncertainty in New Brunswick

In the early 2010s, New Brunswick seemed well positioned to take advantage of the shale gas "revolution" sweeping the United States. New Brunswick had some experience with onshore conventional natural gas extraction, had demonstrated interest from US oil and gas companies in its shale plays, and had voted in a Progressive Conservative government on a platform of responsible development of natural gas. Yet less than five years later the government found itself on defensive, fending off challenges from the opposition that "the Conservative party wants to ... risk your future by gambling on unproven hydro-fracking despite its uncertain economic potential and dangers to drinking water and the environment" (Berry 2014b). On election day in late September 2014, the electorate agreed with the opposition, voting the Progressive Conservative government out of office, along with its proposed framework for shale gas development (Berry 2014a; Logan 2014).

The New Brunswick case demonstrates the ways in which risk narratives can challenge "business-as-usual" development. As a less experienced jurisdiction in comparison to its western counterparts, the New Brunswick government was aware of the need to expand its regulatory frameworks to manage emerging scientific uncertainties related to hydraulic fracturing. However, as government officials and regulators engaged in epistemic learning to manage technical uncertainties, the government's regulatory approach failed to appease the concerns of environmental activists, Indigenous nations, and members of the public, who were doing their own social learning regarding catastrophic harms.

New Brunswick Regulatory Timeline

In September 2010 David Alward's Progressive Conservatives (PC) defeated the incumbent Liberals by a significant margin, winning 42 of a total 55 seats in the legislature (CBC News 2010). In January 2011

Alward established a high level Natural Gas Steering Committee to coordinate the development of new rules as well as a staff-level Natural Gas Group to review regulatory frameworks in other jurisdictions and to draft new rules regarding shale gas (New Brunswick Natural Gas Group 2012b).

In June 2011, in advance of a stakeholder forum on shale gas, the province introduced new rules regarding (1) baseline water testing, (2) chemical disclosure, and (3) security bonds for landowners (G. Bird 2011) and indicated the possibility that it would implement a new royalty regime. In November 2011, the government launched an online "Oil and Natural Gas Map" viewer that included information on well location, name, company, and well depth (*Moncton Times and Transcript* 2011). Minister Northrup also announced in December that the province was developing a new royalty rate for shale gas development based on research in other jurisdictions and indicated that royalty instruments might include an aspect of profit-sharing with landowners (Huras 2011). On 6 December 2011, the Legislative Assembly debated a motion supporting "continued responsible exploration of New Brunswick's Natural Gas reserves in conjunction with the development of a framework of world class regulations to ensure the protection of the residents of New Brunswick, our groundwater, and the environment." Despite attempts by the opposition to declare a moratorium instead, the motion passed (Government of New Brunswick 2012).

In May 2012 the New Brunswick government released the formal discussion paper based on the regulatory review conducted by the Natural Gas Group in 2011 (New Brunswick Natural Gas Group 2012b). Responsible Environmental Management of Oil and Gas Activities in New Brunswick: Recommendations for Public Discussion included 116 regulatory changes that presented a comprehensive regulatory regime designed specifically for shale gas management. The government announced a two-month public consultation tour to be conducted until mid-July 2012 (Huras 2012a). The consultations were headed by Dr Louis LaPierre, an emeritus professor of biology from the University of Moncton (Huras 2012d).

On 15 October 2012 the government released the LaPierre report as well as a health report developed by Chief Medical Officer Eilish Cleary. Both reports were made public despite initial statements by Health Minister Madeleine Dube that Cleary's report would remain private (Chilibeck 2012; Berry 2012). Although both reports documented extensive public concern and lack of public trust regarding hydraulic fracturing, Cleary's report focused on potential detrimental health effects (Berry 2012). By contrast, the LaPierre report promoted the establishment of an independent Energy Institute to provide "peer-reviewed" research to inform policy development (Huras 2012e; LaPierre 2012; Cleary 2012). The government proceeded to establish the Energy Institute and appointed LaPierre as its head in February 2013 (Mazerolle 2013a).

In February 2013, the New Brunswick government released "Responsible Environmental Management of Oil and Natural Gas Activities in New Brunswick: Rules for Industry," a regulatory framework including 97 new rules (Huras 2013a). The framework included regulations on a wide variety of instruments, ranging from seismic testing to well casing, wastewater management, transportation, storage, GHG emissions, and well monitoring and remediation (Government of New Brunswick 2013a; Huras 2013b). The government followed up on the framework in May 2013, releasing its "Oil and Natural Gas Blueprint" (Government of New Brunswick 2013b), which identified six strategic objectives: "1) Environmental Responsibility, 2) Effective Regulation and Enforcement, 3) Community Relations, 4) First Nations Engagement, 5) Stability of Supply, and 6) Economic Development," and corresponding actions. The appendices of the strategy also included direct responses to recommendations of the LaPierre and Cleary reports (Government of New Brunswick 2013b, 2).

The Energy Institute held energy roundtables and energy forums with stakeholders in August 2013. Later in August LaPierre was commended at the US Natural Research Council for pioneering this approach (Mazerolle 2013b; Chilibeck 2013). In a surprising turn of events, in September 2013, CBC Radio Canada investigated LaPierre's scholarly credentials, finding that despite holding a professorship at the University of Moncton for over thirty years, he had misrepresented his academic credentials (CBC News 2013b). LaPierre subsequently resigned from the Energy Institute (Berry 2013).

Shale gas emerged as a prominent issue in the 2014 provincial election, with political observers concluding that Alward's handling of the shale gas political conflict was a factor in the government's loss to the opposition in the general election on 22 September 2014 (Berry 2014a; Logan 2014). Shortly after taking office, the New Brunswick Liberals confirmed that they would implement a moratorium on hydraulic fracturing. In December 2014 Premier Brian Gallant announced a moratorium on all forms of hydraulic fracturing. The moratorium would be lifted once five conditions had been met: (1) a social licence was established through public consultation; (2) baseline information on air, health, and water was gathered and informed the regulatory regime; (3) the government developed a plan for waste water disposal; (4) the government implemented a process to fulfil the province's duty to consult with First Nations; and (5) the government developed an effective royalty regime to garner benefits from production (CBC News 2014).

In 2015 the government appointed a new commission to conduct public hearings on shale gas development in the province (Fast 2016a; New Brunswick Commission on Hydraulic Fracturing 2016). Although the commission stopped short of recommending a moratorium, focusing instead on the need for the province to develop an independent regulator, the government reaffirmed the moratorium in 2016 (McHardie 2016).

The Political Economy of New Brunswick

Although New Brunswick was the site of Canada's earliest natural gas discovery in 1859, by the early 2000s the existing natural gas industry in New Brunswick was modest, providing 1.5 per cent of New Brunswick's GDP (Statistics Canada 2011). Since the 1990s modern onshore production had focused primarily on tight sands oil and gas production using conventional techniques located in the Stoney Creek and McCully gas fields in southern New Brunswick (Rivard et al. 2014; Bott 2004; Government of New Brunswick 2011). As of 2011, New Brunswick had thirty conventional natural gas and sixteen oil wells in operation (Government of New Brunswick 2011). In 2005 Corridor Resources obtained onshore leases for several parcels in Elgin County (Government of New Brunswick n.d.) and conducted a small successful frack. From 2007 to 2009 in conjunction with Apache Canada, Corridor proceeded to fracture two more test wells; however these exploratory drills were less successful in finding commercial reserves (Government of New Brunswick n.d.; CBC News 2011; Hobson 2013c).

In 2009, at an energy conference in Texas, government officials from the Department of Natural Resources were approached by representatives of Southwestern Energy, a major US unconventional gas producer based in Texas (Government of New Brunswick 2009). Southwestern representatives expressed interest in exploring the extent of the New Brunswick shale play, which industry studies had shown contained some of the thickest shale in North America (confidential interviewee 2014b). With changes to global gas markets (Foss 2007), the general investment conditions for unconventional natural gas development had improved, positioning the onshore industry as a potential future revenue stream (Keir 2014).

In addition, by 2009, the New Brunswick government was in need of new sources of investment to address deindustrialization in the north of the province and the need for new power sources, the latter because of challenges in refurbishing the nuclear power plant at Point Lepreu

(Millar Forthcoming; Emery 2020; 2021). In the early 2000s, the province was relatively dependent on coal and oil fuelled turbines to generate electricity, including the supply of Venezuelan Orimulsion for heavy oil power plants in the province (Statistics Canada 2014; Volpé and Thompson 2011). In late 2009 the Graham government attempted to sell the provincial utility, NB Power, to Hydro Quebec for $4.8 billion to reduce the provincial debt, however Graham quickly rescinded the deal in early 2010 in the face of strong opposition from the New Brunswick public, other Atlantic provinces, and industry consumers concerned about changes to electricity rates (CBC News 2009b; Bissett 2010; Macdonald and Lesch 2015).

In this context, investment in shale gas seemed like a potential boon to the province's finances, and in March 2010 the Graham Liberal government granted an exploratory lease to SWN Resources Canada (SWN), a wholly owned subsidiary of Southwestern Energy, for approximately 1.1 million ha, approximately one seventh of New Brunswick's land (Government of New Brunswick n.d.; Stanec Consulting Ltd. 2014). Early discussions between industry representatives and government officials from the Department of Natural Resources, the Department of Environment, and the Department of Energy and Mines established the need for a modernized regulatory framework, but prioritized establishing the extent of the resource first (Keir 2014; confidential interviewee 2014b). These findings suggest that the interests of SWN Canada in exploring the extent of the New Brunswick resource, combined with support from existing onshore producers, helped drive the government's attention toward shale gas development.

In addition to the material interests of industry, the electoral interests of the Progressive Conservatives also played a role in determining the initial pro-development position of the Alward government. Alward had initially raised concerns regarding the potential environmental risks of shale gas development while in opposition. However, by the September 2010 election campaign, the PCs were in full support of "responsible" shale gas development (Alward 2014). The Alward campaign focused on economic growth, with natural gas development as a key component. The PC platform committed to "support[ing] the responsible expansion of the natural gas sector while ensuring the safety and security of homeowners and our groundwater supply" (PCNB 2010). Alward noted that the party's pro-development position was developed through a policy convention and party meetings leading up to the 2010 election, and was spurred in part by pressure from within the party to support a moratorium on uranium exploration (Alward 2014). The platform continued to shape Alward's position, with Alward referring throughout his tenure to the 2010 mandate to justify his government's pro-development position (Gallant 2011).

The Alward government's pursuit of "responsible development" during this time was not internally perceived as contrary to the government's climate commitments. In 2011, building on the 2007 Climate Action Plan, the government released the New Brunswick Energy Blueprint, which affirmed the government's commitment to increasing the provinces' renewable energy target to 40 per cent of electricity generation by 2020 (Government of New Brunswick 2006; NB Department of Energy 2011).

However it is less clear that industry interest groups determined the extent of the regulation as suggested by studies of US states (Richardson et al. 2013). Instead, the New Brunswick government was ahead of industry in determining its regulatory preferences. Although the Canadian Association of Petroleum Producers (CAPP) issued general guiding principles in September 2011 (CAPP 2011), by December 2011 New Brunswick had already released its own principles (New Brunswick Natural Gas Group 2012b), which were more expansive in scope and more detailed than those put forward by CAPP. For example, following the release of the Recommendations for Public Discussion in May 2012, industry representatives argued for less prescriptive regulation, particularly regarding wellbore casings. This preference was not upheld in the 2013 regulatory framework (Huras 2012b; Leonard 2014). Why did New Brunswick eschew the single-issue approach evident in British Columbia in favour of a more comprehensive regulatory framework attuned to environmental harms?

Risk Narratives in New Brunswick Hydraulic Fracturing Debates

Several distinct narratives about the risks of hydraulic fracturing emerged in New Brunswick policy debates (see figures 6.1 and 6.2). Early on, elected officials articulated a pro-development position, lauding the economic potential of shale gas production to address the provinces' ailing economic conditions. For example, early estimates based on production in British Columbia cited potential economic benefits of over $200 million in royalties and 34,000 jobs (Mazerolle 2011), a significant boon to a province with a population of 750,000 residents (Statistics Canada 2018). Industry representatives stressed economic and clean energy benefits to the province but also the proven safety of hydraulic fracturing in western provinces. In an op-ed published in the *Moncton Times and Transcript*, the Canadian Association of Petroleum Producers argued "as illustrated in other Canadian jurisdictions such as B.C., effective regulation and

responsible industry practices ensure pipe and well integrity, and iso-
lation of freshwater sources from contamination" (Pryce 2011). In contrast to narratives of linear risk common in British Columbia however,
both elected officials and industry representatives were more likely to
acknowledge the need to develop a new robust regulatory framework
in the face of potential environmental harms. An early *Moncton Times
and Transcript* article reporting on the position of the Minister of Natural
Resources, Bruce Northrup, illustrates this type of narrative:

> "We feel we've done our homework on this, and we know we have a lot
> of homework left to do," Northrup said, vowing that in the end, this prov-
> ince will have a set of regulations designed to extract the proper amount
> of royalties for New Brunswickers (including landowners) while protect-
> ing the earth, air and water at the same time or there will be no shale-gas
> industry in this province. "If we find we can't do it right, then we just
> won't do it," he said. (Foster 2011b)

Thus, the dominant narrative put forward by elected officials in New
Brunswick was that of complex risk, namely that there were potential
environmental harms to groundwater, but that it was possible for the
government to "get it right" if they consulted with the appropriate
experts. Government officials assured the public that any potential risks
could be addressed through a comprehensive regulatory framework
that would protect New Brunswickers from environmental harm.

In contrast, a broad anti-fracking coalition developed in New Brunswick around catastrophic and uncertain risk narratives. Figure 6.1
shows the sustained high salience of environmental risks in the province
almost throughout the entire period of study. Members of the coalition
included environmental organizations, opposition MLAs, municipal governments, health care professionals, members of Indigenous
nations, and other residents. Figure 6.2 reflects the focus of environmental advocates, Indigenous nations, and other residents on environmental risks, the majority of which concerned fears regarding groundwater
contamination. Initial comments from members of the New Brunswick
Green Party exemplify this catastrophic narrative: in 2010 Green Party
leader Jack McDougall commented: "all that stands between perma-
nently ruining a community's groundwater and the fracking fluids is a
thin sheath consisting of layers of metal pipe and cement that are used
to line the well bore where it passes through the water table ... Fracking is the suspected cause of over 1,000 cases of groundwater contamination across the United States and western Canada" (Weston 2010).
This narrative would later be taken up by environmental advocates and

Figure 6.1. Risk narratives by year in New Brunswick debates

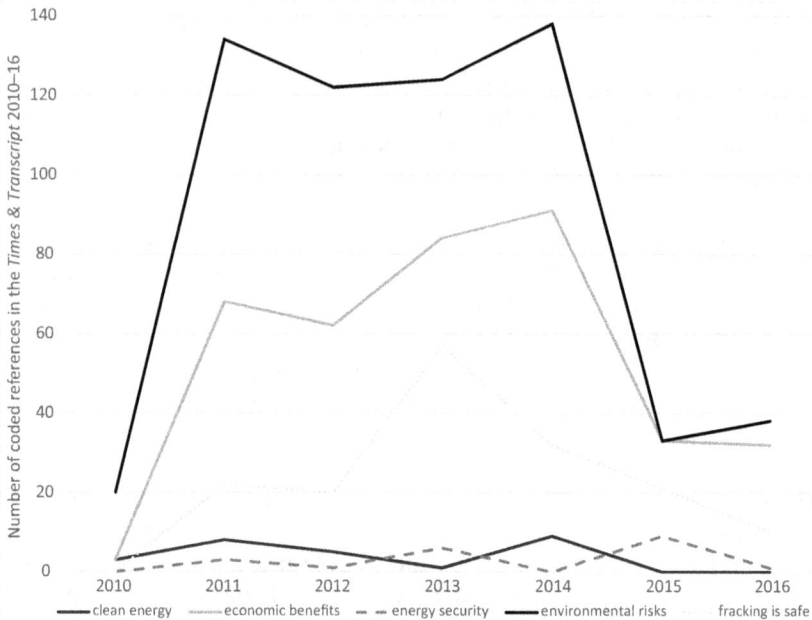

municipal officials concerned about potential contamination of watersheds supplying drinking water to Moncton, the province's largest city
(Lewis 2010; Foster 2010). The intensity of concern regarding groundwater contamination was rooted in dread risk, namely that the contamination of the water supply would be substantial and irreversible. As a
resident noted at an anti-fracking protest in Fredericton in the summer
of 2011: "we just don't want any of this going on. We want the water to
stay the way it is and everything around, the environment to stay clean
for the little ones ... We have to think about the future and not just right
now" (*Moncton Times and Transcript* 2011).

Although less prevalent than discussions of threats to groundwater,
those opposing development also occasionally referred to the economic
risks of development, calling into question the reliability of jobs and oil
and gas markets. Anti-fracking actors also relied on normative claims,
arguing that hydraulic fracturing represented a threat to the rural New
Brunswick way of life. Bureaucrats, journalists, and academics tended
to reiterate arguments that supported and/or opposed development,
resulting in a more balanced distribution overall. These findings suggest
that academics and journalists likely fell on both sides of the debate.[1]

Figure 6.2. Risk narratives by actor type in New Brunswick debates

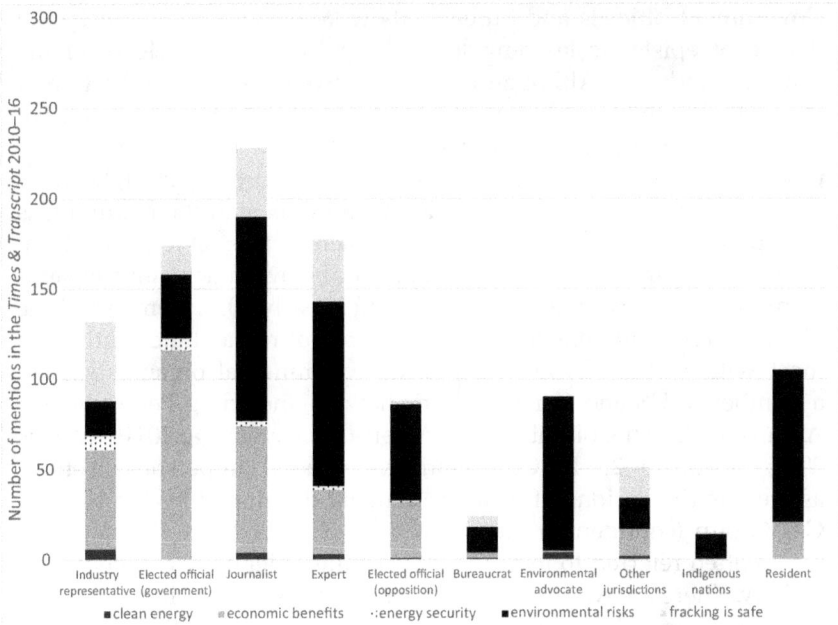

Analysis of the news media dataset thus identifies both complex and catastrophic narratives at play in New Brunswick public debates. Elected officials and industry representatives put forward narratives of complex risk, focusing primarily on economic benefits. Government leaders assured the public that any potential risks could be addressed through robust regulation, which they asserted would be the strongest in North America. On the other hand, anti-fracking actors drew on a battery of unknown and dread risks to call into question the adequacy of the existing regulatory regime. Environmental advocates and Indigenous leaders were likely to call for a ban on hydraulic fracturing, while elected officials from the Liberal and NDP parties were more likely to demonstrate support for an intermediate path through a moratorium. Although narratives of complex risk would shape the regulatory reforms developed by the Alward government through processes of epistemic learning, uncertainty regarding groundwater contamination would inform the precautionary preferences of opposition leaders through a process of social learning, ultimately leading to the implementation of a moratorium once the Liberals took power in 2014.

Complex Risk and Epistemic Learning

Government officials and bureaucrats were engaged in a substantial degree of epistemic learning leading up to the 2013 release of the Rules for Industry. Although the Progressive Conservative Party had committed to the goal of responsible development, in the early days of the mandate the government was uncertain how to best regulate the practice of hydraulic fracturing (Alward 2014; Northrup 2014; Leonard 2014). Both the Natural Gas Steering Committee and the Natural Gas Group included a broad cross-departmental mix of officials from Natural Resources, Environment and Local Government, and Finance (confidential interviewee 2014b). Starting in early 2011, members of the Natural Gas Group embarked on a number of research delegations to meet with regulators, industry, and environmental organizations in a number of US and Canadian jurisdictions, including Pennsylvania, Arkansas, British Columbia, and Alberta (Foster 2011a; 2011b; Morris 2011; Northrup 2014). Consultations included in-person meetings as well as discussions of draft documents developed by the Natural Gas Group (confidential interviewee 2014f; 2014e). Pro-development actors often referred to instances of learning to legitimize government activity. There are twenty-eight references to "learning" in the dataset, the majority of which were from politicians in 2011, as the rules were being developed.[2]

Part of the challenge facing policy makers was that although New Brunswick had a modest onshore industry operating in Sussex and Stoney Creek, the pace of development had been slow and as a result the existing Oil and Natural Gas Act had never been modernized (confidential interviewee 2014f; Leonard 2014; Coon 2014). Rather than develop draft legislation, the experts involved in policy formulation suggested that the group develop a comprehensive set of rules for industry. These rules could be appended to the Environmental Impact Assessment (EIA) regulation, which was enforceable under the Clean Environment Act (confidential interviewee 2014f; Environment and Local Government 2012). The institutional context thus provided support for a stronger environmental approach among the members of the Natural Gas Group, a position that was also backed by political direction from above (confidential interviewee 2014f). The bureaucrats involved in drafting the rules had a strong mandate to increase the stringency of the framework to be "the best in North America" (confidential interviewee 2014f; 2014b; 2014e). Although New Brunswick does not have an independent Oil and Gas Commission as in British Columbia, evidence suggests that the institutional position of the Natural Gas Group served to somewhat insulate actors from external interference, at least in the beginning of policy development. As in British Columbia, the institutional context served to facilitate a process of epistemic learning among bureaucrats, arguably a more extensive and in-depth process than the limited technical learning within the BC Oil and Gas Commission.

The Recommendations for Public Discussion released in 2012 drew on a wide range of regulatory frameworks from other jurisdictions, scientific research and policy reports issued by both industry associations and environmental groups[3] (New Brunswick Natural Gas Group 2012a). Although the recommendations drew heavily on Alberta regulatory frameworks in a number of places with regard to well construction and casing (New Brunswick Natural Gas Group 2012b), the drafters of the report also relied considerably on emerging studies from New York examining air quality and fracturing fluid mobility (confidential interviewee 2014f; 2014e; 2014b). Despite the wealth of information gathered, interviewees noted the paucity of peer reviewed scientific information available in 2011, stressing the lack of available information on health risks as well as best practices for public consultation (confidential interviewee 2014f; 2014e; 2014b; Cleary 2012; confidential interviewee 2014c). The breadth and depth of policy research undertaken by members of the Natural Gas Group suggests that government actors did not simply adopt the preferences of industry representatives when drafting the New Brunswick regulatory framework. Indeed, the Department of Environment and Local Government developed an annotated bibliography of over one thousand policy documents and scientific research articles on shale gas that it shared through an informal network with bureaucrats in other jurisdictions (Department of Environmental and Local Government, NB 2014; confidential interviewee 2014g). Interview data and document analysis suggest an indirect route of industry influence, with New Brunswick actors learning from available scientific evidence as well as from regulators who were guided by the policy legacy of oil and gas development in their own jurisdictions.

The Natural Gas Group's extensive consultation with regulators from other jurisdictions, and the depth and breadth of scientific research conducted suggests that bureaucrats in New Brunswick engaged in a substantive process of epistemic learning that led to changes in both their causal beliefs about the policy problem as well as the means through which the problem should be addressed. From the experiences of other jurisdictions, civil servants came to believe that the risks of hydraulic fracturing were manageable and could be regulated. As one interviewee described:

> It was interesting, because I went from "this is interesting, I understand
> the science behind it," to seeing Gasland and thinking "this is horrible,
> it's going to destroy our province," to then going back to really studying
> it. I remember ... being very much opposed, and then the scientist in me
> kicked in and I started looking and digging and was like "okay no, not so
> bad, so how do we do this right?" (confidential interviewee 2014e)

Beliefs within the civil service thus shifted from a slightly uncertain
position in 2010 regarding policy instruments to a certain stance in 2013
regarding the feasibility of the draft rules. This shift is echoed in the
rhetoric that was used by politicians in the media to promote the new
rules. Throughout 2012 and 2013, elected officials of the Alward government increasingly referred to New Brunswick's regulatory framework
as the "strongest in North America," to a peak of twenty-one references
in the *Moncton Times and Transcript* in 2013. The regulatory approach put
forward by the Natural Gas Group was institutionalized and embedded
in the establishment of the Department of Energy and Mines in 2012,
when the hydraulic fracturing file was transferred from the Department
of Natural Resources to the new leadership of Energy Minister Craig
Leonard. As a result of visiting regulators in other jurisdictions, Leonard was strongly convinced that the technology of hydraulic fracturing
was proven to be safe, and saw the rules as a key mechanism for assuag-
ing public concerns (Leonard 2014).

As beliefs regarding regulation solidified among pro-development
actors, they also began to rely on appeals to expert authority to legitimize
the new policy direction. Elected officials stressed the capacity of science
to manage risk, often asserting that evidence-based policy making would
resolve rising public conflict. Importantly, both elected officials and
industry actors repeatedly referred to problems of "misinformation" to
discredit anti-fracking claims, asserting that the New Brunswick government's sources were scientific and credible. This position was institutionalized through the establishment of the New Brunswick Energy Institute
(NBEI) in the fall of 2013. Arising from LaPierre's recommendations from
the 2012 summer "listening tour" (LaPierre 2012), the NBEI was designed
to foster research on the specific conditions of shale gas development
in New Brunswick (confidential interviewee 2014a). Both Leonard and
Alward saw the NBEI as a mechanism to dampen public opinion and
bolster the legitimacy of government through the credibility of science
(Alward 2014; Leonard 2014). Leonard notes:

> The information was getting so tainted, whether it was government say-
> ing certain things, industry saying things, those opposed making their

points, that it got to the point where the average person was simply saying "I don't know who to believe anymore" ... So the concept was, let's go to the one area that still has credibility in our society, which is the academics, they are supposed to be unbiased, well trained, and experts in their fields and we'll set up this institute. We'll recruit these individuals and obviously there will be research dollars available. We'll have these individuals go do work, and do the research on these pressing questions that people have, and as a result of that hopefully we'll be able to better educate people on what is taking place. If we do determine that there are issues then we can address them, or if the research indicates that there are no issues, then hopefully that would give people more confidence to move forward. (Leonard 2014)

Despite the perceptions of politicians that the NBEI would dampen political conflict, the success of the NBEI was highly contingent on the credibility of the scientists involved. The termination of LaPierre in the fall of 2013 seriously undermined the legitimacy of the nascent organization in the face of the public, a weakness which was exacerbated by the Opposition. Commenting on the future of the NBEI in late 2013, Liberal leader Brian Gallant argued that the NBEI was primarily a rhetorical exercise to promote shale gas development, noting "With the past chair going through his difficulties, many people in New Brunswick and throughout North America will question the credibility of any of the data and any analysis done by the Energy Institute" (Huras 2013d). Ultimately, the NBEI was unable to muster sufficient attention to shift the public debate (Leonard 2014; Alward 2014; Northrup 2014; confidential interviewee 2014a).

The New Brunswick case suggests that epistemic learning can lead to more comprehensive regulatory change if bureaucratic actors are given a degree of discretion within the existing institutional context. Given a free reign to develop new rules within a relatively autonomous institutional and legislative framework, the members of the Natural Gas Group approached the problem of how to manage the uncertainty of hydraulic fracturing in a comprehensive way, focusing on a variety of instruments to regulate seismic testing, well construction, chemical disclosure, wastewater storage and disposal, and land use, among other potential risks.

In chapter 3, I posited that epistemic learning is likely facilitated by narratives of complex risk. To reiterate, complex risk narratives are those which stress the complexity of causal ideas, making it difficult for actors to predict the extent of potential environmental harm. Greater frequency of complex risk narratives in media debates suggests a situation in which political actors are relatively certain of their overall policy objectives but are less sure of the instruments necessary to convert potential environmental harms into tolerable or acceptable risks. In keeping with scholarship on knowledge utilization (Blyth 2001; Haas 1992; Lindvall 2009), I argue that when politicians are functioning from a perception of complex risk they are more likely to turn to substantive bureaucratic or academic expertise in order to guide policy change. Findings from the New Brunswick case provide tentative support for this proposition. Unlike their counterparts in the Opposition, David Alward, Craig Leonard, and Bruce Northrup rarely articulated concerns about scientific challenges to the viability of the shale gas industry, instead maintaining their campaign commitment to "responsible development." Rather than using expertise to deal with uncertainty regarding policy objectives (Blyth 2001; Haas 1992), New Brunswick politicians drew on bureaucratic expertise to resolve the problem of how best to develop shale gas – that is, what policy instruments to adopt. The push toward responsible development was thus translated by bureaucratic actors into a comprehensive and stringent regulatory regime built through a process of epistemic learning from academic experts and regulators from other Canadian and US jurisdictions. The New Brunswick case thus provides additional support for the theoretical claim that expert ideas are more likely to influence instrument choice than they are broader policy objectives (Lindvall 2009; Rietig 2018).

Importantly, however, the organizational process of epistemic learning within the bureaucracy regarding instruments also led to changes in politicians' beliefs regarding political strategies. Both the premier and the energy minister came to believe that turning to scientific expertise and drawing on expert authority would boost the government's legitimacy and resolve rising political conflict (Alward 2014; Leonard 2014). Ultimately, this position left the government open to attacks from the anti-fracking movement regarding the credibility of science and by extension the credibility of the state writ large. The following section examines how anti-fracking actors were able to facilitate a series of authority contests to expand the scope of conflict beyond the confines of the Department of Energy and Mines and to push for a precautionary approach to shale gas development.

Uncertain Risk and Social Learning

Analysis of the New Brunswick case suggests that the anti-fracking movement used a series of focusing events to direct public attention to the environmental and health risks of hydraulic fracturing, calling into question the accuracy of the narratives put forward by pro-development actors. By the time the Alward Conservatives formed the government in September 2010, reports of well failures and contaminated water incidents in the United States had begun to surface. These incidents were encapsulated in Josh Fox's documentary *Gasland* which had premiered at the Sundance Film Festival in January 2010 (Fox 2010).[4] Together with the knowledge that SWN had acquired exploration rights to approximately one seventh of New Brunswick's land, residents began to voice concerns regarding the potential environmental risks facing New Brunswickers (Northrup 2014). Media analysis demonstrates a rapid rise in the salience of hydraulic fracturing in the *Moncton Times and Transcript*; the number of articles mentioning the technology increased sharply from ten mentions in 2010 to eighty-seven in 2011.

One of the early points of contact for people concerned about hydraulic fracturing was the Conservation Council of New Brunswick (CCNB). Under the leadership of David Coon, who would later become the leader of the provincial Green Party, the CCNB had been researching the policy area, focusing on air and water contamination (Coon 2014). By spring 2010 the CCNB had added a strategic program in water protection policy and it released an initial policy primer in the summer (CCNB 2010). Throughout 2010 the volume of calls from concerned residents continued to increase and by the fall of 2010 the environmental organization had begun to host community meetings throughout New Brunswick on the issue (confidential interviewee 2014d; Coon 2014). The CCNB continued to respond to inquiries from communities across the province throughout 2011, facilitating information flow between emerging grassroots groups and organizations (confidential interviewee 2014d; Coon 2014). Although the CCNB's campaign was primarily "home grown," staff drew on the resources of some environmental groups working on hydraulic fracturing in New York state; representatives from some New York groups participated in a few community meetings in Penobsquis and Elgin County in the Fall of 2010 (confidential interviewee 2014d).

At the same time, there was a growing perception among New Brunswick residents that the Alward government was not adequately consulting the public on the shale gas file. Figures 6.3 and 6.4 show the distribution of references to unknown and dread risks across time and by actor types. Anti-fracking actors voiced numerous concerns regarding the relatively closed consultation process, the limited degree of public debate and dialogue, and an overall lack of transparency. These concerns were exacerbated by the initial forum hosted by the Natural Gas Group in June 2011. Although environmental advocates were invited to the table, the general public was not, fuelling

Figure 6.3. Dread and unknown risks by year in New Brunswick debates

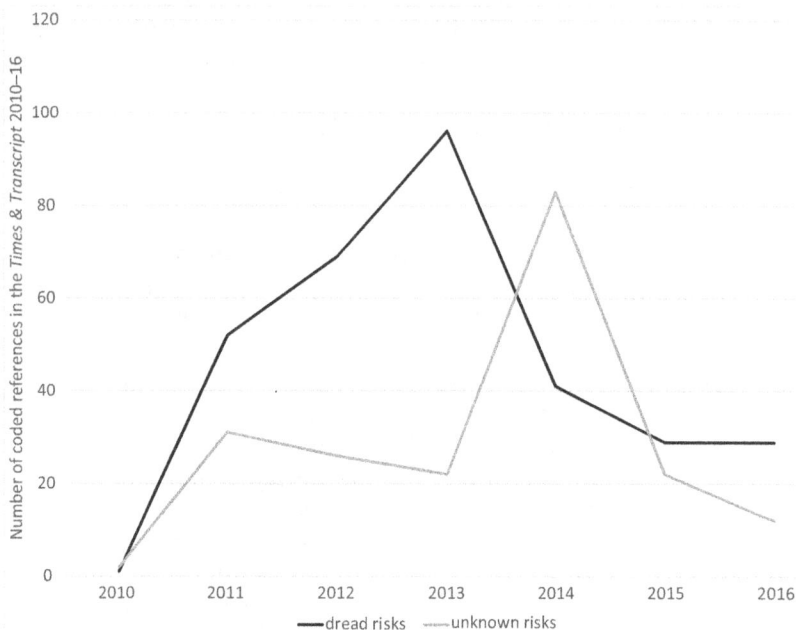

perceptions that the meetings were being held in private and prompt-ing a small protest rally outside of the forum (G. Bird 2011). As one observer notes, "[Government actors] were just doing so much behind closed doors that the general public didn't see a role for themselves in the discourse ... There wasn't a formal mechanism for public engage-ment and it wasn't until much later that they really figured this out" (confidential interviewee 2014d).

The sense of public distrust in government was fuelled in part by previous policy failures on environmental issues that were still reso-nant among New Brunswick residents. Since the late 1990s, residents near the potash mine in Sussex had been experiencing problems with well water contamination. Following an initial environmental assess-ment conducted by the Department of Environment indicating the possibility that the potash mine was the source of contamination, in 2010 the affected residents brought an action before the New Bruns-wick Commissioner of Mines, resulting in a long and protracted series of public hearings (Clancy 2013; Howe 2015; confidential interviewee 2014e). Beyond water concerns, a significant number of New Brunswick residents had also mobilized in 2009 and 2010 against the sale of New

Brunswick Power to Quebec, voicing fears that the Government of New Brunswick was not attentive to citizens' needs or interests (Macdonald and Lesch 2015). These early events served to feed into a general sense of malaise and discontent among the public, evident in the comments of one resident who was quoted in a *Moncton Times and Transcript* article as saying "in my 50 years, I've never felt so disengaged and disenfranchised by a government, ever" (Huras 2012c).

Interviewees from both pro-development and anti-fracking sides of the hydraulic fracturing debate identify the 2012 public consultation headed by Louis LaPierre as a key focusing event that served to open up the scope of conflict (Alward 2014; Leonard 2014; Coon 2014). From the beginning, the "tour" was taxed with a somewhat conflicted mandate. On the one hand, government actors were aware that the Recommendations for Discussion reflected a significant degree of technical complexity and were unlikely to be substantially modified following public input (confidential interviewee 2014e; 2014b). On the other hand, negative public opinion had escalated throughout 2011 and the government was aware of the need to respond by engaging the public in policy design. The resultant "listening tour" placed government actors in the uneasy position of presenting and requesting feedback on technical information without engaging in the relative merits of shale gas development as a policy goal (Leonard 2014; confidential interviewee 2014b). This made the panel particularly vulnerable to attacks on the legitimacy of the entire consultation process, spurring numerous complaints concerning the lack of government integrity (LaPierre 2012). As one observer noted,

> Ninety-nine per cent of the people who got up and spoke were opposed to the industry ... it was amazing – people got up with papers, academic papers, and they had the lines highlighted that they wanted to read, because no one had had a platform to say what they knew about or what they had learnt ... it was too late in the game to have an honest, open, frank conversation before people got to the end of their rope, before they were so frustrated, before you couldn't say anything that would change their minds (confidential interviewee 2014d).

The 2012 public consultations thus became a key venue for anti-fracking advocates both to widen the debate and to challenge the legitimacy of the government's pro-development priorities. Anti-fracking actors shifted the debate from a targeted discussion regarding the viability of various policy instruments to a broader debate regarding the environmental harms of shale gas development. Beginning in public meetings in 2011 and 2012, and continuing through to protests in 2013 and 2014, anti-fracking speakers raised environmental, health, and economic risks connected to hydraulic fracturing. While groundwater and drinking water risks were the most salient throughout the period, additional concerns emerged in 2012 regarding seismic activity and climate change risks due to GHG emissions. In 2013, following the release of the Cleary report, concerns voiced in the media regarding health risks also began to climb. The multifaceted nature of anti-fracking causal beliefs served to facilitate ties among a broad range of actors, sustaining the issue in the public eye. An interviewee noted:

> I think there was something for everybody ... if you're an environmentalist, you're concerned about climate change, if you're a rural landowner, you're worried about property rights ... if you're a homeowner, you're concerned about potential water problems ... even if you are a promoter of economic development and your community is drying up, a lot of people saw through the jobs argument really quickly ... the jobs are transient and the same people move around to where there is work. It wasn't a solid enough argument, even on the economics front. (confidential interviewee 2014d)

The shift of the debate to uncertain environmental harms further intensified during the summer of 2013 as anti-fracking protesters, led by members of the Elsipogtog First Nation, began to target SWN's seismic testing in Kent County (Toogood 2013). Contrary to assertions from industry that seismic testing posed minimal environmental risks (Hobson 2013a), protesters and residents began to refer to the need to steward the land and protect groundwater from potential harms, reflecting a more precautionary approach (Hobson 2013b). Following the October standoff in Elsipogtog, media debates continued to refer to the potential for hydraulic fracturing to undermine the quality of life in rural New Brunswick. Although economic benefits and environmental risks continued to dominate debate, anti-fracking activists were able to widen the debate to incorporate a few other concerns, including the lack of potential economic benefits.

The strength of the anti-fracking movement was bolstered by two parallel but complementary narratives regarding risk that can be elucidated by the typologies of catastrophic and uncertain risks described in chapter 3. In contrast to government actors operating within a narrative of complex risk, residents and some environmental advocates tended to focus on the high probability of dread risks (see figure 6.4). Anti-fracking actors drew on narratives of irreversible environmental harms

Figure 6.4. Dread and unknown risks by actor type in New Brunswick debates

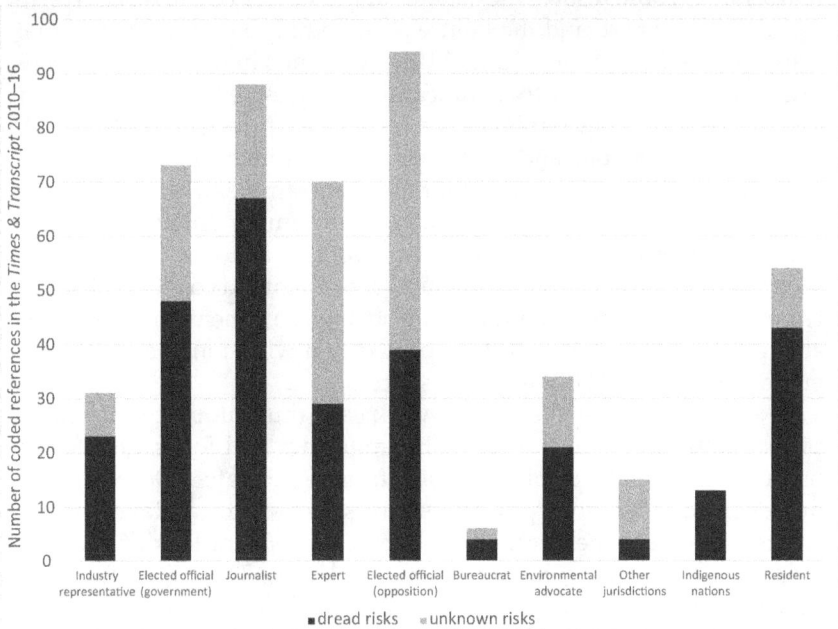

to lend weight to the need for a ban, asserting that it would be impossible to protect against the detrimental impacts of hydraulic fracturing (Huras 2013c). The high proportion of claims in the media regarding the probability of drinking water and/or groundwater risks supports Sunstein's (2005) hypothesis that dread risk narratives can capture public attention, heightening the salience of an issue and ultimately demanding increased public attention.

At the same time, elected officials in opposition were much more likely to focus on unknown risks or scientific uncertainty to justify a temporary moratorium. In a striking use of scientific discourse to support his position, Brian Gallant deliberately referred to current studies being conducted by the Canadian Council of Academies and the US Environmental Protection Agency in the media and the NB Legislative Assembly. In his response to the throne speech in December 2013, he went so far as to read two paragraphs of the EPA progress report to justify a more precautionary approach (Government of New Brunswick 2013c). By focusing on scientific uncertainty – namely, the difficulty experts were experiencing in accurately estimating the probability of potential harms – Gallant was able to undermine the narrative that current science would resolve the debate.[5] Together with the loss of LaPierre as a credible expert, and the failure of the NBEI to provide substantial leadership in the debate (Huras 2013e), Gallant's highlighting of scientific uncertainty served to significantly undermine the government's assertions that the technology was tested and proven safe. Although narratives of uncertain and catastrophic risks ultimately reflected a contradiction with regard to probabilities – for example, whether the risks of hydraulic fracturing are inevitable or are simply unknown – in the short-term, both narratives implicitly supported a moratorium as a preferred policy solution. The agreement among a broad coalition of actors regarding means provided the anti-fracking movement with considerable power. The risk narrative appealed to a wide range of voters, as evident in Gallant's 2014 electoral success.

The findings of the New Brunswick case thus support the general proposition that "policy losers" (Baumgartner and Jones 1993) often use new venues to catalyse public framing debates. Despite commandeering substantive power resources, pro-shale gas development actors were nevertheless hampered by a significant lack of capacity regarding public consultation on energy issues. To address this gap, bureaucrats and politicians relied on scientific expertise, a shift that was ultimately vulnerable to challenges to scientific credibility and the legitimacy of the policy-making process writ large. It is plausible that, in the absence of the scandal regarding LaPierre's academic credentials, the New Brunswick Energy Institute might have been able to bolster the legitimacy of the government's pro-development position, however it is also possible that the NBEI would have functioned as an additional venue in which anti-fracking advocates could contest the proposed regulatory reforms.

In contrast, the anti-fracking movement, although limited by coordination challenges in linking a broad base of actors, was nevertheless able to harness the institutional opening provided by the public consultation process to generate a sustained level of issue salience. Anti-fracking actors were able to take advantage of the 2012 shale gas hearings to attract the public's attention to the dread risks of hydraulic fracturing. By attacking both the legitimacy of the consultation process and the scientific credibility of government officials, anti-fracking actors were able to undermine the perceived viability of the government's pro-development stance among the electorate. Gallant's Liberals were able to exploit this vulnerability throughout the 2014 election campaign, ultimately resulting in their implementation of a moratorium soon after forming government.

In comparison with events in British Columbia, the New Brunswick case illustrates the ways in which a strong industry presence does not necessarily correspond with single-issue regulation. Although the New Brunswick government was elected on a pro-development platform and initially faced relatively low levels of environmental mobilization, in the absence of an independent regulator government officials were less insulated than their western counterparts from electoral politics. At the same time, the higher degree of epistemic or scientific uncertainty regarding the technical complexity of "how to get it right" resulted in a narrative of complex risk, rather than the narrative of linear risk that dominated in British Columbia. As a result, although both provinces demonstrate processes of policy learning, in British Columbia the process was primarily technical, drawing on the internal expertise within the BC Oil and Gas Commission, while in New Brunswick the process relied more heavily on epistemic input from scientific experts and regulators from other jurisdictions. The resulting New Brunswick regulation was thus much more comprehensive than British Columbia's, as regulators aimed to develop a framework for the entire industrial process of hydraulic fracturing in one package.

Moreover, the New Brunswick case demonstrates the ways in which the opening of institutional structures, even through a temporary review process, can interact with uncertain and catastrophic risk narratives to trigger processes of social learning among elites, leading to more precautionary regulation. These dynamics were repeated through the subsequent more targeted consultations led by the government-appointed New Brunswick Commission on Hydraulic Fracturing in 2015. Again, while industry representatives stressed the potential economic benefits of the practice and the overall safety of the industry, members of the public, civil society groups, and Indigenous nations expressed significant concerns regarding dread risks, particularly impacts to groundwater (Fast 2016a; Nourallah 2023). In their 2016 report the commission noted ongoing public concern regarding environmental harms, lack of trust in government to regulate the harms, and more broadly an increased desire for public engagement in policy decision making regarding energy development in the province (New Brunswick Commission on Hydraulic Fracturing 2016). Following the release of the report, in 2016 the provincial government reaffirmed its commitment to a moratorium on shale gas development, reiterating the five conditions that would need to be resolved prior to lifting the moratorium, namely, (1) indication of social licence from the general public; (2) "clear and credible information" regarding environmental harms: (3) plans to address infrastructure impacts; (4) establishment of processes regarding the duty to consult with Indigenous Nations; and (5) a clear royalty framework for development (Department of Energy and Mines 2016). To justify the ongoing moratorium, the Gallant government continued to draw on the twin narratives of scientific uncertainty regarding produced water risks and the lack of trust in government, both among the public and particularly regarding the need for the comprehensive engagement of Indigenous nationss in the province. The made-in-New Brunswick precautionary approach demonstrates the endurance of uncertain and catastrophic risk narratives in the Maritime provinces in contrast to their western counterparts and the ways in which these risk narratives continue to reinforce processes of social learning.

Chapter Seven

Contesting Uncertainty in Nova Scotia

On 3 September 2014, a little over a week after the release of a report on hydraulic fracturing from an independent review, Energy Minister Andrew Younger announced that the Government of Nova Scotia would introduce legislation in the 2014 fall session "to prohibit the use of hydraulic fracturing in shale oil and gas projects" (Younger 2014a). Replacing the de facto moratorium which had been implemented in 2012 by the previous Nova Scotia NDP government, the ban reflected the most stringent regulatory response to hydraulic fracturing to date in Canada. It placed the government of Nova Scotia firmly within a precautionary approach. Heralded by environmental organizations and local community groups as a step in the right direction, it was derided by industry representatives and the federal finance minister (*Halifax Chronicle Herald* Editorial Board 2014). To some observers the decision to implement a ban reflected a political calculus, despite the energy minister's assertion that the legislation was justified based on the lack of scientific certainty respecting the safety of the practice (Younger 2014a).

While the British Columbia case demonstrates the conditions leading to single-issue regulation and the New Brunswick case illustrates the dynamics behind comprehensive regulation and moratoria, the Nova Scotia case presents an opportunity to examine why and how governments engage in more precautionary regulatory approaches. This chapter examines the empirical puzzle of why Nova Scotia initially implemented a moratorium on hydraulic fracturing earlier than the neighbouring jurisdiction of New Brunswick, as well as why the NS government subsequently chose to implement a more stringent ban in 2014.

The Nova Scotia case illustrates that in the context of a relatively nascent oil and gas industry, agitation from a strong local environmental grassroots movement in Nova Scotia spurred the government's initial foray into technical learning to guide regulatory development.

But because of electoral concerns, the provincial government quickly implemented a temporary moratorium to halt development citing the need for an external review, which reflects similar dynamics to the New Brunswick case. Over time, the de facto moratorium sustained the salience of both unknown and dread risks in public debates, establishing a policy position from which it was difficult for provincial government to retreat. This chapter argues that although the government commissioned a substantive independent review into scientific evidence, the degree of scientific uncertainty on the issue was a necessary but not sufficient cause to prompt the government's decision to implement its ban on high volume hydraulic fracturing. Instead, I find the external review directed the attention of government officials to narratives of catastrophic risk.

Nova Scotia Regulatory Timeline

In June 2009, the Nova Scotia NDP, led by Darrell Dexter, won thirty-one seats in the provincial election, defeating the incumbent Progressive Conservatives (CBC News 2009a). For the first few years of its mandate, the NDP government focused primarily on developing a plan to reach its campaign target of 25 per cent renewable energy by 2015, building on the work of the previous government set out by the Environmental Goals and Sustainable Prosperity Act and the 2009 Energy Strategy (NSNDP 2009; M. Adams, Wheeler, and Woolston 2011; Government of Nova Scotia 2007; Nova Scotia Department of Energy 2009). Nevertheless, in early 2011 Energy Minister Charlie Parker began to meet with officials in other jurisdictions to explore potential regulatory frameworks for shale gas development, travelling to Oklahoma to meet with the Interstate Oil and Gas Compact Commission, the Oklahoma Corporation Commission, and the Groundwater Protection Council (Baxter 2011). In April 2011, along with Environment Minister Sterling Belliveau, Parker announced that the government was launching an internal review on hydraulic fracturing, to be carried out by representatives of the two departments (Government of Nova Scotia 2011a).

A year later the government announced a two year "hold" on hydraulic fracturing to study the policy issue, formally extending the internal review process and indicating that it would not approve new applications from oil and gas producers to engage in hydraulic fracturing in the interim (Government of Nova Scotia 2012b; Canadian Press 2012a).

During this time period, the Department of Environment was also engaged in managing the storage and remediation of produced waste water from three wells fracked by oil and gas company Triangle

Petroleum in 2007 and 2008, which was being stored in two ponds in Hants County near Kennetcook (Bundale 2011b; 2011a). Wrangling between Triangle Petroleum and the Department of the Environment regarding the waste water ponds continued throughout the fall, primarily due to the discovery of naturally occurring radioactive materials (NORMs) in the water (Canadian Press 2012b; MacDonald 2012). By November 2012, Triangle had missed the government's deadline to clean up wastewater ponds, citing new scientific and technical issues, and noting that while Triangle had originally anticipated using deep injection to deal with the waste water, the government had rejected this option as geologically inviable (Bundale 2012b).

The Nova Scotia government continued to face challenges in dealing with fracturing waste water in other municipalities as well, as Atlantic Industrial Services attempted to attain a permit to discharge treated produced water from fracked New Brunswick wells into the Colchester municipal water system (Gorman 2013a; 2012). Following a discovery that Atlantic Industrial Services had also discharged fracturing fluids through the Windsor sewage system without testing the water for NORMs, some Colchester county residents appealed the council's decision to discharge wastewater (Gorman 2013b). The Colchester Appeals Committee overturned the council decision to discharge water through the sewage system in May 2013, leaving Atlantic Industrial without a mechanism to disperse the waste water (Beswick 2013). In August 2013, a few months before the general election, the Dexter NDP government commissioned a new independent external review of hydraulic fracturing, and appointed Cape Breton University President David Wheeler to lead it (Gorman 2013d; Government of Nova Scotia 2013a).

Despite the attention garnered by the independent review, hydraulic fracturing was not a particularly salient issue during the election, with all three parties supporting the moratorium despite small differences that made the Liberals' platform slightly more pro-environment than the NDP and PCs (Davene 2013; Ross 2013). In October 2013 the NDP lost to the Liberals by a landslide, with the Liberals winning thirty-three seats and the NDP falling to seven (CBC News 2013c). In December, the new Liberal government passed the Importation of Hydraulic Fracturing Wastewater Prohibition Act preventing import of fracking waste water from other jurisdictions (Gorman 2013e; Government of Nova Scotia 2013b).

Beginning in February 2014, the nine-member expert external review committee, headed by David Wheeler, began to release bi-monthly discussion papers focusing on different aspects of hydraulic fracturing such as drilling, wastewater remediation, and economic modelling of costs and benefits based on gas in place estimates (Ross 2014b). Each draft chapter was "published" online and available for comment and review on a two-week rolling deadline; in addition, town hall meetings were conducted in July in ten different communities throughout the province, headed by Wheeler and other members of the panel (Ross 2014c; Ayers 2014a). In August 2014 Wheeler released the report, recommending a slow, incremental approach to policy development in full consultation with municipalities and Indigenous nations (Gorman 2014b).

In September 2014 Energy Minister Andrew Younger announced the governments' intention to introduce legislation prohibiting fracking during the fall session of the legislature (Erskine 2014). On 14 November, Bill 6, An Act to Amend Chapter 342 of the Revised Statutes, 1989, the Petroleum Resources Act passed third reading in the House; the amendment prohibits "high-volume hydraulic fracturing in shale unless exempted by regulation for the purpose of testing or research" (Government of Nova Scotia 2014b).

The Political Economy of Nova Scotia

From a broad vantage point, the oil and gas industry in Nova Scotia is more established than in New Brunswick, raising the possibility that oil and gas interests could have influenced provincial politicians' regulatory preferences. In 2009, Nova Scotia's energy mix was substantially dependent on fossil fuels, with 89 per cent of electricity generation dependent on conventional steam and internal combustion; New Brunswick's mix was slightly more varied with hydro-electric power contributing 22 per cent in comparison to Nova Scotia's 9 per cent (Statistics Canada 2014). Since the early 1990s Nova Scotia's oil and gas production has been based on offshore development, first with the offshore Cohasset-Panuke oil field, which began production in 1992, followed by the Sable Offshore Energy Project in 1999 (CCEI 2007). Since Cohasset-Panuke finished oil production in 2002, several companies have considered drilling deeper for natural gas and liquids; Encana developed the Deep Panuke project which produced gas from 2013 to 2018 (ICF Consulting Canada 2013; Taylor 2013). Industry's perspective on the Nova Scotia regulatory regime was that it was generally favourable towards investment during this time period; a 2010 survey conducted by the Fraser Institute found that petroleum industry executives ranked Nova Scotia in the relatively attractive second quintile, just behind British Columbia and ahead of Alberta in terms of pro-development regimes (Angevine and Cervantes 2010, 29).

The provinces' extensive experience with offshore drilling has not, however, translated into a strong onshore oil and gas presence. To date, only a few companies have signed leases for conventional oil and coalbed methane exploration and production (Natural Resources Canada 2015) and only three onshore gas wells have been hydraulically fracked in the Horton Group shale near Windsor-Kennetcook (Taylor 2012; Canadian Press 2012b; Rivard et al. 2012; ALLconsulting 2012). Exploration was conducted by Elmworth Energy, a subsidiary of Triangle Petroleum, a company headquartered in Denver with experience working in the Bakken shale in North Dakota (Taylor 2012; Triangle Petroleum Corporation 2015). The exploration wells were vertically fracked but did not yield a commercial discovery (Natural Resources Canada 2015; Nova Scotia Fracking Resource and Action Coalition and Harris 2013; Nova Scotia Department of Energy 2012). Following exploration, Elmworth/ Triangle stored the produced water in two large holding ponds near Kennetcook, where the water remained until 2016 (Nova Scotia Fracking Resource and Action Coalition and Harris 2013; *Halifax Chronicle Herald 2014*; Summers 2016).

Apart from Triangle, interest in onshore exploration in Nova Scotia has been minimal, with a few other minor companies engaged in production (Bundale 2012a; Demont 2014). Because of limited exploration, quantitative estimates of the size of the unconventional resource are uncertain, ranging from 17 to 69 Tcf (gas in place) (US EIA 2013b; Atherton et al. 2014). Since 2008 there has been little indication of any major oil and gas companies resuming exploration of Nova Scotian resources; by and large, the major offshore oil and gas companies have maintained a low profile in public debates, despite the potential for offshore hydraulic fracturing (Bundale 2011c).

Beyond oil and gas, successive provincial governments during this period focused their attention on developing the renewable energy industry in Nova Scotia. In 2007 the Progressive Conservative government under Rodney Macdonald passed the Environmental Goals and Sustainable Prosperity Act, which established quantitative targets for renewable energy production in the province; prior to the election in 2009 the PCs followed up with Toward a Greener Future, an energy strategy outlining a firm commitment to increase renewable electricity energy to 25 per cent by 2020 (Nova Scotia Department of Energy 2009). Once in government, the NDP, which had also campaigned on a 25 per cent renewable target (NSNDP 2009), spent the early part of its mandate focusing on developing the Renewable Electricity Plan, which it released in 2010 (Nova Scotia Department of Energy 2010). The plan focused on wind, solar, biomass, and tidal energy as potential sources for energy production and outlined mechanisms for both large and community projects to feed into the grid. As a part of plan development, the NDP government also contracted a consulting team led by

Wheeler to conduct public consultations in the fall of 2009; Wheeler had previously facilitated a consultation on the Progressive Conservatives' energy efficiency strategy in 2008 (Nova Scotia Department of Energy 2010; M. Adams, Wheeler, and Woolston 2011).

In the latter part of 2010 the Nova Scotia government was also engaged in extensive consultation with the Government of Newfoundland for access to hydro-electricity produced at Muskrat Falls (Macdonald and Lesch 2015). The negotiated agreement between Nalcor Energy and Emera, parent company of Nova Scotia Power, entitled Nova Scotia to 170 mw annually at a firm rate for thirty-five years, in return for a $6.2 billion investment toward the building of a sub-sea cable between Cape Ray Newfoundland and Cape Breton, as well as other grid improvements (Macdonald and Lesch 2015; Premier's Office 2010). The deal provided baseload power to Nova Scotia to complement intermittent production driven by other in-province renewables (Premier's Office 2010).

These characteristics of Nova Scotia's energy mix suggest that despite the more visible presence of fossil fuel production in Nova Scotia, the relative influence of onshore oil and gas companies respecting unconventional gas development was weaker in Nova Scotia than in New Brunswick during the period of study. In the absence of any major interest from industry (such as the SWN lease in New Brunswick), hydraulic fracturing development was not high on the government agenda, leaving case management and regulation to the line departments of Energy and Environment (Black 2014). Although the economic potential for unconventional development was appealing, the NDP cabinet was focused primarily on ramping up the renewable plan and marketing offshore production, rather than pursuing onshore investment (Black 2014). A subsequent federal shipbuilding contract in 2012 also provided the Dexter government with additional revenue, reducing the immediacy of the need to pursue shale gas for economic benefits (Demont 2012; Black 2014).

Another potential reason for the more limited influence of economic incentives on the Dexter government's policy position on shale gas was the partisan preferences of the cabinet. Arguably, the Nova Scotia NDP had stronger connections to the environmental lobby in Atlantic Canada than the New Brunswick PCs (Black 2014; confidential interviewee 2014h), and presumably as a more left leaning party, environmental concerns also play more strongly to the NDP base. However, the environment was not a salient issue in the 2009 election. In her analysis of the 2009 election, Turnbull (2009) stresses that ideological differences between the NDP and the PCs were in fact minimal during the campaign. Turnbull argues that Dexter's choice to leave Howard

Epstein, an experienced MLA and former head of the Ecology Action Centre (EAC), out of cabinet reflected a strategy to position the NDP as centre-left party, signifying a movement away from environmental concerns. The lack of evidence for a strong partisan/ideological influence on party positions on hydraulic fracturing also holds true for the 2013 election, where there was very little disagreement across party platforms (Ross 2013). These findings suggest that in contrast to New Brunswick, where partisan politics may have had a strategic influence on determining David Alward's preferences, the influence of deep ideological beliefs regarding oil and gas development was less definitive in the Nova Scotia case. As in New Brunswick, understanding the scope and form of Nova Scotia's regulatory response to hydraulic fracturing requires closer attention to the role of risk narratives in facilitating attention to potential environmental harms.

Risk Narratives in Nova Scotia Hydraulic Fracturing Debates

Several characteristics distinguish Nova Scotia media debates from the other provinces (see figures 7.1 and 7.2). First, although government officials waded into public debates very early on in New Brunswick and British Columbia, in Nova Scotia the NDP government had a quiet media presence on hydraulic fracturing, with only a few public statements by Energy Minister Charlie Parker on the purpose of the internal review (Bundale 2011a) and Environment Minister Stirling Belliveau regarding disposal options for the Kennetcook wastewater (Zaccagna 2012). Figure 7.1 demonstrates that the bulk of arguments attributable to elected officials were made after the 2013 election, focusing on the independent review and the ban on imported wastewater. There were also very few statements made by elected officials in opposition, from either the Liberals or the Progressive Conservatives. In contrast to New Brunswick, where politicians led the pro-development coalition, elected officials in Nova Scotia were more likely to mediate between the relative safety of the technology on the one hand and environmental risks on the other in their public statements, reflecting a more centrist position.

By and large, industry leaders in Nova Scotia focused on the proven safety of hydraulic fracturing. Representatives from Atlantic firms Triangle Petroleum and Corridor Resources stressed that "hydraulic fracturing has been used for decades on millions of wells with no incident" (Bundale 2011c). In contrast to Albertan oil and gas CEOs, the Atlantic companies were more likely to dismiss environmental complaints as completely unfounded, with one representative commenting in the *Halifax Chronicle Herald* that "what is asserted without proof can be

Figure 7.1. Risk narratives by year in Nova Scotia debates

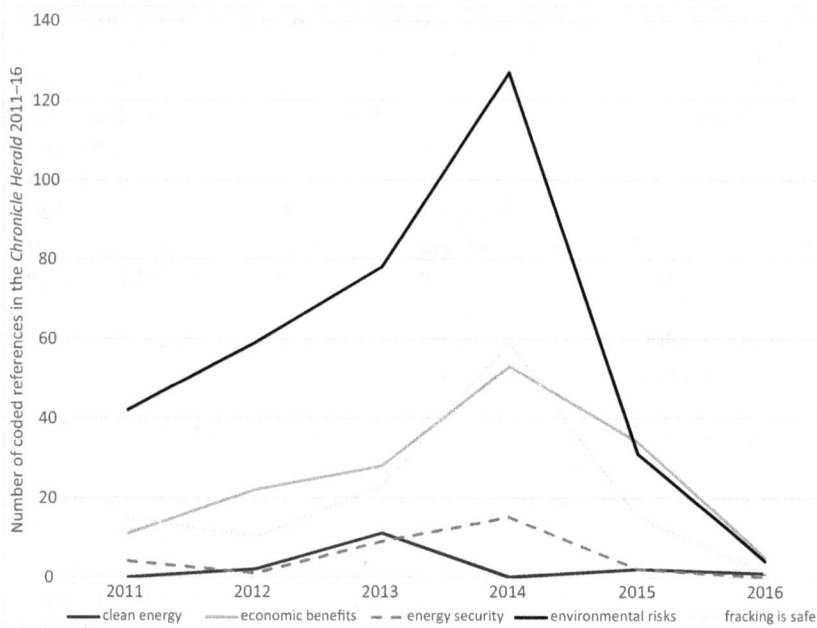

Figure 7.2. Risk narratives by actor type in Nova Scotia debates

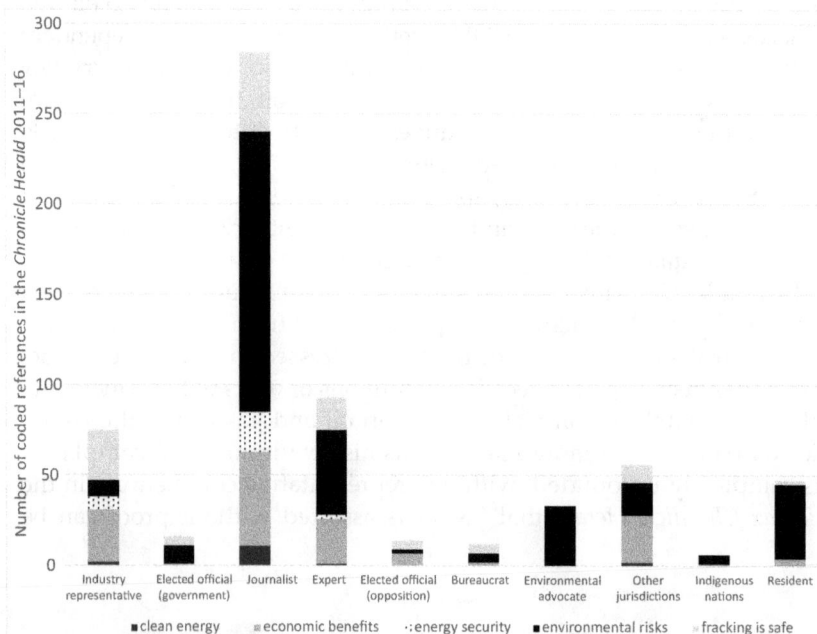

dismissed without proof ... These so-called environmentalists are paid just to oppose shale gas development. They are not interested in truth or facts. They are just interested in opposing and obstructing" (Bundale 2011a). Despite this resolute stance, figure 7.2 demonstrates that the total frequency of industry commentary was lower than in Alberta or New Brunswick, illustrating the more diffuse influence of the oil and gas industry in Nova Scotia.

In the absence of strong signals from either elected officials or industry representatives, the early actors on the scene in media debates were journalists themselves, with significantly fewer actor arguments attributable in 2011 and 2012 to specific industry leaders, academics, residents, or environmental advocates in the dataset.[1] However, after the announcement of the independent review in 2013, there was a substantial spike in public arguments made by academics and technical experts as well as an increase in claims made by residents. Journalists were more likely to focus on environmental risks, covering groundwater risks in both Nova Scotia and internationally. As in Alberta, the Halifax Chronicle Herald media articles referred consistently to policy positions and environmental debates occurring in other jurisdictions: over the four year time period the study coded references to policy makers in eighteen different jurisdictions, including a range of US states, New Brunswick, Quebec, Newfoundland, France, the European Union, and the United Kingdom.[2] In aggregate, references to other jurisdictions split between economic benefits on the one hand and environmental risks on the other, with some mention of the relative safety of the practice.[3] Together, the less polarized and more multifaceted coverage in Nova Scotia reflects an ambiguous, somewhat undefined risk narrative among elected officials and bureaucrats, which included elements of both linear and uncertain risks.

Conversely, environmental advocates, Indigenous community members, and other residents often articulated concerns regarding environmental harms to justify their opposition to hydraulic fracturing. As in New Brunswick, concerns regarding potential catastrophic risks to groundwater were prominent, as were other air and land impacts, with environmental advocates noting "the big issues we are concerned with are water consumption and contamination, air quality, and way of life" (Bundale 2011a). Despite some support for development to bring jobs to the region (Beswick 2012), local residents continued to expressed concerns around groundwater contamination especially in Inverness County (near PetroWorth's conventional well) and Colchester County, where fracking wastewater was being processed (MacIntyre 2013; Gorman 2013c). Nevertheless, the overall salience of environmental risks in the dataset was driven by journalists reporting on purported environmental harms, rather than specific claims put forward by individual Nova Scotian activists.

The Nova Scotian media debates thus reflect a more ambiguous, somewhat linear risk narrative among pro-development advocates, countered by both uncertain and catastrophic narratives regarding groundwater risks. Industry arguments in the Nova Scotia dataset were much less bullish on the economic benefits of hydraulic fracturing than in New Brunswick and tended to rely more extensively on claims regarding the relative safety of the practice in other jurisdictions, including Alberta and British Columbia, to counter calls for prohibition. At the other end of the spectrum, environmental advocates and residents were vocal regarding water contamination and water management risks. In the middle of the debate were elected officials, journalists, and academics, articulating the classic trade-off between the economic benefits of hydraulic fracturing on the one hand and environmental harms on the other. The tentative linear risk narrative set the stage for an early foray into limited technical learning within the civil service, which was interrupted by the de facto moratorium. Yet a gradual, intensifying narrative of catastrophic risk among activated publics in the province eventually triggered a rapid process of political learning among elected officials, resulting in the ban.

Uncertain Risk and Limited Technical Learning

In comparison to the Alward government in New Brunswick, the Dexter NDP government faced weaker economic incentives to move quickly to develop a comprehensive regulatory framework for hydraulic fracturing. The absence of a strong onshore oil and gas lobby, together with alternative mega-projects for revenue generation such as the federal shipbuilding contract and the government's prior commitment to upscaling the renewable industry, positioned hydraulic fracturing as a low priority on the government agenda (Black 2014). Nevertheless, throughout 2010, bureaucratic staff in the energy and environment ministries had begun to conduct informal internal research on hydraulic fracturing regulations in other Canadian jurisdictions (confidential interviewee 2014g; 2014i).

During 2010 hydraulic fracturing began to climb in salience with the Nova Scotia public, primarily in response to increased coverage of US environmental conflicts and harms, both in the mainstream and social media. As in New Brunswick, environmental advocates began to receive a volume of calls from concerned residents looking for more information regarding the practice of hydraulic fracturing and potential environmental impacts; by the fall of 2010 both the Sierra Club and the Ecology Action Centre (EAC) had directed staff attention to the issue (Black 2014; confidential interviewee 2014k; 2014h; Minkow 2015). Sitting MLAs in the NDP government had also begun to field calls and inquiries with respect to fracking. These MLAs included Charlie Parker, MLA for Pictou West, who would be appointed Minister of Energy as well as Minister of Natural Resources in January 2011.[4] By the end of 2010 the umbrella group, the Nova Scotia Fracking Action and Resource Coalition (NOFRAC), was established, representing twelve member organizations and over fifty individual members (MacDonald 2011; Nova Scotia Fracking Resource and Action Coalition n.d.; Minkow 2015). In March 2011 the Sierra Club continued to publicly pressure the government to act, calling for a ban on hydraulic fracturing, particularly as it related to PetroWorth's conventional oil and gas projects in Lake Ainslie (Sierra Club Atlantic Chapter 2011). The potential of electoral risks raised political concerns within the party caucus that the government needed to do something on the issue (Black 2014). However, what the goal of government action should be, and by extension, what shape the regulatory response should take, was indeterminate.

In April 2011 the government announced that it would embark on the internal review led by a joint team of bureaucrats from the Departments of Energy and the Environment (Government of Nova Scotia 2011a). Interviews with key actors in the policy subsystem confirm that the mandate of the review was understood to be constrained to a technical review of potential environmental impacts and a regulatory review of different responses in a variety of jurisdictions. In distinction from the New Brunswick internal review, policy actors reported that there was more uncertainty both at the political and bureaucratic level regarding the broader goal or purpose of the review (confidential interviewee 2014g; 2014h; 2014k). Although New Brunswick bureaucrats had been directed to develop the "best regulation in North America" so that New Brunswick was well prepared to engage in hydraulic fracturing, bureaucrats in Nova Scotia were much less clear whether the industry would be developed at all. As one interviewee noted,

> The government at the time was of two minds about the process. It was an NDP government, so there was a strong environmental bent, but at the same time they also wanted to do some business development, but this wasn't an existing business; the offshore was quite big but the onshore had never really launched. So they found themselves in this real grey area, so they wanted us to do some work, but not to offend anybody. (confidential interviewee 2014g)

As a result, the informal internal review process in Nova Scotia was less insulated from electoral demands than in the Natural Gas Group in New Brunswick. The unspoken assumption for many government officials was that the internal review was intended to neutralize the potential for public debate (Black 2014; confidential interviewee 2014g; 2014i). Both bureaucrats and government officials hoped to contain the debate within a technical risk analysis, assessing potential environmental impacts and identifying best management practices for chemical disclosure, well construction, and waste management (Government of Nova Scotia 2011b). The review team was explicitly tasked not to engage in general public consultation (confidential interviewee 2014i); however, after internal discussion, the team did solicit online and written commentary on the scope of the internal review in June 2011 (Government of Nova Scotia 2011b). These findings suggest that institutional insularity in Nova Scotia was substantially weaker than in New Brunswick or British Columbia, resulting in a less robust process of technical learning.

The seven member internal review team, comprising co-leads from the energy and environment ministries as well as three experts from energy and four from environment (Government of Nova Scotia 2011c), conducted its review throughout the summer and winter of 2011, consulting with representatives from Alberta, British Columbia, New Brunswick, Quebec, Texas, Ohio, and Wyoming (Precht and Dempster 2012). In comparison with the New Brunswick Natural Gas Group, the review team spent fewer resources travelling to other jurisdictions, although the Department of Energy drew on its connections within the US Interstate Oil and Gas Compact Commission to identify potential contacts and informants for the review (confidential interviewee 2014i). The team also contracted two consultants, Paul Precht and Don Dempster, to conduct the jurisdictional regulatory scan. The resulting report was released in March 2012 and provided one of the first comparative analyses of provincial regulatory frameworks in both Atlantic and western Canada.[5] Team members engaged in both negative and positive lesson drawing, noting the limited capacity of assessment staff in Wyoming, who reported conducting more than 10,000 well approvals annually, as well as the comprehensive scope of the Alberta regulatory system for conventional oil and gas (confidential interviewee 2014g; 2014i). The work of the review team focused mostly on the experiences of other regulators, rather than findings from the scientific literature, which interviewees noted was nascent and underdeveloped at the time of the review (confidential interviewee 2014g). Despite these advances, the scope of work conducted by the Nova Scotia internal review team seems to be less comprehensive than that of their New Brunswick counterparts.

Organizational and administrative factors also undermined the team's policy capacity. On an administrative level, as a joint initiative between the Departments of Energy and of Environment, the team had to negotiate historic boundaries between the two, evident in the perception that although the Department of Environment had a greater wealth of technical expertise, the Department of Energy controlled the budget (confidential interviewee 2014g; Black 2014). Beyond problems of coordination, it was difficult for the team to find a "home" for any of the potential regulations or best practices identified by the scan that could be adapted to the Nova Scotia context. As in New Brunswick, the legislative framework for onshore oil and gas development in Nova Scotia is quite limited. As a result, much of the work is managed on a case-by-case basis, with civil servants empowered with a high degree of discretion (confidential interviewee 2014g; Wheeler et al. 2015). Although regulatory discretion had been adequate in addressing well applications to date, bureaucrats were aware that discretionary models can be problematic in the eyes of the public. Together these factors limited the team's findings being communicated up to the political level and out to the public. One interviewee commented on the challenges in communicating review findings:

> [The internal review identified a] huge difference between the Canadian and US regulatory regime. And I think with the whole broad public fear spectrum, much of the fear was driven by US instances. But the whole regulatory regime here is different. So it was very challenging dealing with perceptions that had already been formed, many of them on things that would never occur in Canada ... the review team had thought that that might be the case, but it was reinforced really strongly during those interviews [with regulators from other jurisdictions]. The challenge was that the review team didn't have a mandate to go out there and do much talking about that. (confidential interviewee 2014i)

Limited technical learning at the bureaucratic level was counterbalanced by increasing external signals at the political level that the public was agitating for a moratorium or ban. Throughout 2011 the Sierra Club and the Ecology Action Centre had continued to call for a ban on high volume hydraulic fracturing (Bundale 2011c). Advocates pointed out that a high proportion of responses to the public consultation on the scope of the internal review (46 per cent) had mentioned the need for a ban or a moratorium (confidential interviewee 2014h; Government of Nova Scotia 2011d). On the ground, NOFRAC had continued to mobilize, organizing a range of public actions, including lobbying MLAs' constituency offices and bringing in scientific experts to speak on potential environmental risks

(Minkow 2015; confidential interviewee 2014k; Gorman 2011). At the same time, other members of NOFRAC continued to pressure the government behind the scenes, meeting with bureaucrats and ministers on both the environmental and energy files (confidential interviewee 2014k).

Growing public pressure and unease was a key factor in the NDP's decision to announce the two-year hold in April 2012. By and large, the internal review had not identified increasing scientific uncertainty as a key element of the policy context, nor had the line departments recommended a de facto moratorium (confidential interviewee 2014i; 2014g). Although the NDP cabinet had considered the need to wait for scientific consensus in the form of the Canadian Council of Academies report or the US EPA report (Black 2014), the decision to announce the two-year hold was more likely a politically expedient decision designed to dampen public anxiety without scaring off investment by legislating a moratorium, a distinction that was arguably lost in the press (Black 2014). Importantly, although the need for more extensive research was cited by the government as the rationale for extending the two-year hold (Government of Nova Scotia 2012a), an unintended effect of the decision was that the bureaucratic team stopped work on the issue and did not release its comprehensive report on the internal review. In effect, the announcement of the moratorium dampened the internal capacity of the bureaucracy to engage in technical learning. This was contrary to the process in New Brunswick, in which the work of the Natural Gas Group was insulated early on from external pressures and then institutionalized in the policy direction of the newly established Ministry of Energy.

Catastrophic Risk and Political Learning

Although policy decision makers anticipated that the de facto moratorium would dampen public attention to hydraulic fracturing, throughout 2012, environmental advocates and local anti-fracking groups continued to raise concerns regarding the efficacy and legitimacy of the internal review. Their efforts were aided in part by ongoing disputes between Triangle Petroleum, the parent company responsible for two holding ponds of produced water from the Kennetcook frack in 2007 and 2008, and the Nova Scotia Department of Environment regarding the appropriate method of disposal for the ponds (MacDonald 2012). In 2011, the Ecology Action Centre filed freedom of information requests regarding the wells in Hants County. Members of NOFRAC spent two years analysing the data, eventually releasing a report in April 2013 (Nova Scotia Fracking Resource and Action Coalition and Harris 2013). The NOFRAC report commended the Nova Scotia Department of Environment for its refusal to approve multiple requests from Triangle to use deep well injection to dispose of wastewater, a practice that is common in the Bakken shale but untested in the Horton Group Shale. Nevertheless, the report identified several regulatory gaps in the Department of Environment's assessment process, including concerns with water withdrawal approvals, enforcement of site reclamation, and soil testing. Most important, the report brought to light the government's lack of experience in testing for various chemical and radioactive elements potentially present in the Kennetcook ponds and documented several questionable approvals for wastewater disposal, including allowing Triangle to dispose of frozen wastewater on site and Atlantic Industrial Services to discharge water (originating from Kennetcook) through the Windsor Sewage Treatment Plant. In particular, the Windsor case highlighted communication problems between the Department of Environment and Windsor City Council, the latter whose members were unaware of the potential for increased contaminants when they approved the discharge of water through the treatment plant (Nova Scotia Fracking Resource and Action Coalition and Harris 2013). The NOFRAC report concluded by calling for a ten-year moratorium or outright ban on hydraulic fracturing in Nova Scotia.

Throughout the spring of 2013, the problem of produced water disposal also began to resonate with municipal actors. Colchester County blocked Atlantic Industrial Services' attempt to dispose produced water imported from New Brunswick (in addition to other water from the Kennetcook frack) (Gorman 2013c; Beswick 2013) and Inverness County banned hydraulic fracturing altogether. Although municipalities lack legal standing to enforce a ban on resource development since the latter is the purview of the provinces, discussions in council and town hall meetings served to raise the salience of the issue both among community members and among the wider public. Similar to mobilization dynamics in New Brunswick, public concerns regarding hydraulic fracturing built upon perceived previous failures of the Department of Environment to protect the public against industrial pollution, including the Sydney tar ponds and more recently water pollution from the pulp and paper industry in Boat Harbour (confidential interviewee 2014h; 2014i; 2014k; Wheeler et al. 2015; Waldron 2018). Together with the Department of Environment's "step-by-step, well-by-well, permit-by-permit" regulatory approach, these failures of legitimacy exacerbated levels of mistrust among the public, whose concern was that the government was failing to regulate hydraulic fracturing in their interest (Nova Scotia Fracking Resource and Action Coalition and Harris 2013).

The notion that the public's support for hydraulic fracturing was tentative at best was confirmed by survey research, undertaken by Corporate

Research Associates in May 2013, which found that 53 per cent of Nova Scotia residents opposed shale gas development (Alberstat 2013). The survey findings echoed the national poll conducted by Environics in 2012. These dynamics clearly affected perceptions within the NDP government, with a number of politicians concerned about the impact of the policy on their electoral chances in the upcoming 2013 election (Black 2014). At the same time, political staff were aware of the dangers of enacting a hasty policy decision in advance of other jurisdictions, as well as the political risk in making hydraulic fracturing a campaign issue since there was very little distinction between the parties on the issue and thus a lack of a comparative advantage (Gorman 2013d; Black 2014).

Within this context, the government considered establishing an independent external review headed by Dr David Wheeler, the President of Cape Breton University. Wheeler was known on both the political and bureaucratic levels in Nova Scotia because of his work leading consultations on the energy efficiency strategy and the renewable energy plan in 2008 and 2009 respectively. From a political perspective, delegating responsibility for the review to an external body would inoculate cabinet from potential electoral risks (Black 2014). From a bureaucratic perspective, the external review also addressed concerns raised during the internal review regarding trust deficits among the public, especially since internal research indicated that the academic community was more likely to engender greater levels of trust (confidential interviewee 2014i). The NDP government formally announced the independent review in August 2013, prior to the kickoff of the election campaign. The decision was criticized by both the Liberals and PCs as way to punt the issue past the October 2013 election (Gorman 2013d). Although the announcement was supported by environmental groups in the province, hydraulic fracturing was a non-issue in the election campaign, as were environmental issues generally (Davene 2013; Ross 2013).

Once in power, the Nova Scotia Liberals, led by Premier Stephen McNeil, confirmed their support for the mandate of the independent review. A panel of experts was commissioned to conduct public consultations on hydraulic fracturing and develop a literature review of potential risks. Unlike the internal review, the independent review included health and socio-economic impacts (Wheeler et al. 2015; Atherton et al. 2014). As discussed above, Wheeler was experienced in designing and guiding public consultations on energy issues in Nova Scotia. He included a range of mechanisms to incorporate public input into the review, including soliciting suggestions for panel members' skill sets and panel nominees, inviting general submissions to the panel, and gathering written feedback on specific discussion papers (Wheeler et al.

2015). The review was designed to include several iterative cycles of input centred on the staggered online release of ten discussion papers. The discussion papers covered a range of aspects such as potential economic benefits, health risks, socio-economic effects, well integrity and environmental impacts (Wheeler et al. 2015). After each discussion paper was released, the public had two to three weeks to submit commentary back to the panel. Public input was then incorporated into the final draft under the discretion of the expert who drafted the chapter (Wheeler et al. 2015; Ross 2014b). Beyond written input, the panel held two informational meetings on the general design of the review, three online forums on specific topics related to the discussion papers, and eleven public meetings across the province on the full draft report and tentative recommendations, which took place in the summer of 2014 (Wheeler et al. 2015; Atherton et al. 2014). In contrast to the New Brunswick "listening tour" headed by Dr LaPierre, the Nova Scotia independent review was decidedly precautionary in its tenor. The final report identifies a precautionary approach as a key guiding principle for the panel and defines precautionary in reference to the UN Conference in Environment and Development, the Nova Scotia Environmental Goals and Sustainable Prosperity Act (2007), and the Mi'kmaw concept of Netuklimk (Atherton et al. 2014). Prior to the release of the discussion papers, Wheeler asserted that the "quasi-academic" scope of the review provided the panel with greater flexibility to include a wide range of inputs, from technical expertise to traditional Indigenous knowledge (Ross 2014b; 2014a). As such, the independent review reflected an intent to open up the policy process and engage the public in decision making.

The review received a volume of unique written submissions on general aspects of the review (238) as well as the discussion papers (170). By and large, the majority of respondents identified as residents rather than representatives of organized interest groups and were in support of a moratorium or ban on the practice (Wheeler et al. 2015; Atherton et al. 2014). The release of the discussion papers also corresponded with a spike in media debates regarding various potential risks. Figure 7.1 shows the distribution of coded references to environmental risks in the *Halifax Chronicle Herald* over time. Although groundwater and drinking water risk frames were salient throughout the period from 2011 to 2014, in 2014, there is a significant growth in media references to alternative risks such as health impacts, climate change risks, and land-use concerns in the dataset.

Alongside claims of environmental harms, media coverage during the independent review also reflects a spike in mentions of scientific uncertainty; figure 7.3 documents dread and unknown risks over time in the dataset. A dominant theme from both the media analysis and the final

Figure 7.3. Dread and unknown risks by year in Nova Scotia debates

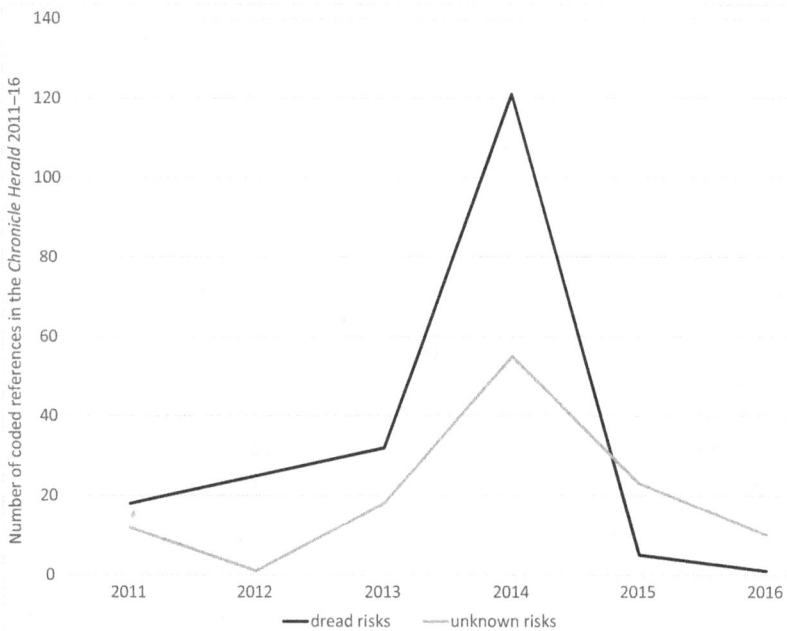

report of the independent panel was that the state of scientific knowledge regarding hydraulic fracturing was uncertain and that more research was needed to inform government action. In parallel with media debates in New Brunswick, actors asserted that scientific analyses had the tools to calculate risks; however, most actors appealing to expert authority in Nova Scotia were journalists and academics, rather than government actors, as was the case in New Brunswick. A smaller subset of claims also asserted that science has proven that hydraulic fracturing poses a low probability of environmental harm. However, in Nova Scotia those claims were countered by other actors who stressed that the science was contested; residents also questioned the source of findings (e.g., industry funded studies versus independent studies). Overall, the media debates during the independent review reflect a greater concern with the vagaries of scientific knowledge than in New Brunswick.

Despite the panel's extensive efforts to engage the public in the independent review, there were growing concerns among environmental advocates and local community groups that the panel was highly technocratic and top-down in its approach (confidential interviewee 2014h;

2014k). Environmental advocates and other academics were concerned that the panellists had not incorporated the extensive scientific information provided to the panel through the submissions process into the draft discussion papers. Environmental groups were disappointed in the first few reports that examined hydraulic fracturing processes, resource potential, and groundwater impacts (confidential interviewee 2014h; 2014k). After the independent panel's release of the groundwater discussion paper, the Ecology Action Centre expressed concerns publicly that the panel's reports reflected an industry bias and lacked scientific rigour (MacDonald 2014). Advocates' frustration with the perceived lack of receptivity of the panel culminated in five different anti-fracking groups (the Council of Canadians, Powershift Canada, LeadNow, the Ecology Action Centre, and the Atlantic Chapter of the Sierra Club) coordinating their efforts through a joint organizing team. Dubbed the "league of ladies," it sought to mobilize the public to attend the eleven town hall meetings in the summer of 2014 (Minkow 2015; confidential interviewee 2014k). In particular, the organizing team put together tool kits for residents attending meetings to bolster the confidence of speakers in commandeering scientific rationales in support of their arguments against fracking (confidential interviewee 2014k; Minkow 2015).

As a result of the mobilizing efforts of environmental advocates and grassroots organizers, the Nova Scotia public meetings in the summer of 2014 became an ad hoc venue within which anti-fracking advocates were able to reframe the debate and to challenge the legitimacy of the independent review, and by extension, the government. Despite the panel's original intent to facilitate dialogue between opposing views (Wheeler et al. 2015), the majority of the public meetings were dominated by anti-fracking narratives that wove together concerns regarding the scientific credibility of the panellists and general apprehensions regarding the trustworthiness of the government to regulate environmental harms (Ayers 2014a; F. Campbell 2014; Gorman 2014a; Delaney 2014). Figure 7.3 illustrates the steady rise in arguments in the *Halifax Chronicle Herald* in 2014 referring to dread risks, including the need for public dialogue, lack of trust in government, and public opposition to hydraulic fracturing. As in New Brunswick, anti-fracking advocates attacked the scientific credibility of the panellists to call into question the overall legitimacy of the independent review. Early critiques of the discussion papers argued that the reports did not include available peer-reviewed evidence (MacDonald 2014), while later arguments levied attacks against the scientific autonomy of Wheeler himself (Ayers 2014b).

The public meetings became a temporary institutional venue through which anti-fracking activists were able to contest the authority and

Figure 7.4. Dread and unknown risks by actor type in Nova Scotia debates

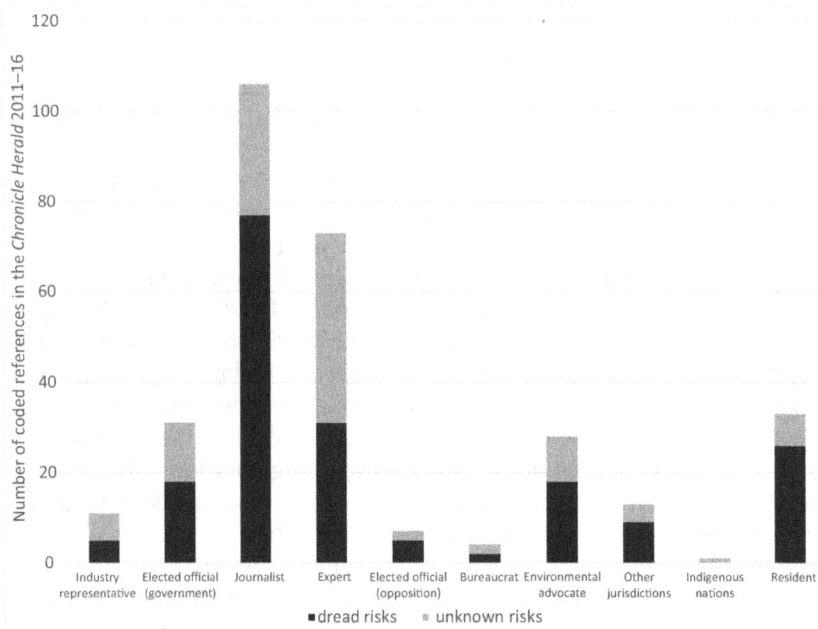

legitimacy of the government to regulate and develop policy regarding hydraulic fracturing and shale gas development in Nova Scotia. From a political perspective, the independent review failed considerably in its implicit mandate to contain public contestation. Indeed, as the media analysis demonstrates, the public meetings served to exacerbate, rather than dampen the volatility and salience of public debate.[6] Nevertheless, unlike the LaPierre consultation in New Brunswick, the independent panel was able to retain a degree of credibility and standing among journalists, bureaucrats, and government officials, although it is difficult to assess whether its resilience was because of Wheeler's ability to withstand the critique on his personal credentials or whether the design of the consultation provided the panel with a cache of procedural legitimacy that was difficult to undermine (Schmidt 2013). Yet there was a tension in the mandate of the Nova Scotia review between the desire to air (and presumably neutralize) public complaints on the one hand and engage in technical research on instruments on the other, which was difficult for experts to reconcile. Despite the extensive research into consultation design (Wheeler et al. 2015), the panel, and in particular

Wheeler, were vulnerable to procedural critiques from environmental advocates suggesting that the panel was rooted in a technocratic, academic, and top-down approach that precluded any genuine discussion of policy goals. At the same time, the independent review confirmed that there was considerable scientific uncertainty within existing literature regarding the objective risks of hydraulic fracturing (Atherton et al. 2014). The final report of the panel attempted to thread the needle between technical risk assessment and summarizing public opinion by acknowledging that the province was unable to accurately assess the risks of hydraulic fracturing, and that it was not the right time for rapid development, although the panel stopped short of recommending an outright moratorium (Atherton et al. 2014; Gorman 2014b). Nevertheless, shortly after the release of the report, Energy Minister Andrew Younger announced that the government would legislate a prohibition on the practice of high-volume hydraulic fracturing in Nova Scotia (Younger 2014a).

The conditions identified above reflect processes of social learning driven by a narrative of uncertainty (e.g., high unknown and high dread risk). The Liberals' decision to impose a ban on hydraulic fracturing after the release of the Wheeler report could be the result of processes of social learning in which the dominant actors came to change their preferences regarding the core construction of the policy problem – that is, that hydraulic fracturing is an inherently risky industrial process with potentially catastrophic and irreversible harms to the environment. In his justification of the ban both in the media and on his personal blog (Younger 2014a; 2014c; 2014b), Younger referred to the degree of scientific uncertainty surrounding the practice, noting:

> Across North America, the debate over hydraulic fracturing has turned to dueling documentaries, and studies, all claiming to hold the absolute truth about the safety and merits of using hydraulic fracturing. The Council of Canadian Academies, in its report for the government of Canada, summed up the reason for such a split in opinions stating: "Many of the pertinent questions are hard to answer objectively and scientifically, either for lack of data, for lack of publicly available data, or due to divergent interpretations of existing data." (Younger 2014b)

However, the speed and magnitude of Younger's response to the Wheeler report suggests that by the time the report was released, a catastrophic narrative rooted in dread risk had taken hold, triggering processes of political learning. The two key arguments used by government officials (and later journalists) to justify the ban were that (1) the government had listened to Nova Scotians and (2) as a result they found that they were not ready for development. Younger provided a clear example of this type of argument during the press conference announcing the ban, arguing that "Nova Scotians have put their trust in our government, that we will listen to the concerns and not allow a process that most Nova Scotians are just simply not comfortable with at this time" (Erskine 2014).

Younger's defence of the ban in the House also reflects a predominant focus on the lack of social licence among Nova Scotians for hydraulic fracturing, referring to public opposition from concerned residents, Mi'kmaw nations, and the Union of Nova Scotia Municipalities (Government of Nova Scotia 2014b). In contrast to New Brunswick's Premier Gallant, who relied predominantly on arguments regarding scientific uncertainty to justify his policy position, Younger was much more likely to rely on popular authority to legitimize his actions.[7] The Nova Scotia case suggests that signals regarding electoral unrest guided the attention of Younger and the Liberal caucus away from social learning towards a more political calculation. Together with municipal bans and consultations with Mi'kmaw nations, the town hall meetings demonstrated a significant degree of public discontent regarding dread risks that the Liberals were unwilling to exacerbate, especially without the guarantee of economic benefits. Data also suggest that Younger was monitoring events in New Brunswick and was aware of both Gallant's position and the potential pitfalls of public unrest (Black 2014). Following the release of the Wheeler report, Younger commented to the *Moncton Times and Transcript* "What I've learned from the New Brunswick model and from the history here is that I have a responsibility to show people we are listening to them ... I'm concerned about how this issue is tearing certain communities apart" (Morris 2014).

Thus, although the Nova Scotia case provided the opportunity for belief change among regulators and elected officials, the path toward the ban on high-volume hydraulic fracturing was more likely the result of political learning spurred by catastrophic risk narratives. Unlike environmental advocates or residents voicing their concerns at public meetings, Younger was personally less concerned about dread risks. Indeed, his initial announcement of the ban refers to the probability that the practice can be regulated safely, as in western Canada (Younger 2014b). Nevertheless, Younger was keenly attentive to the dynamics of public opposition to hydraulic fracturing. His remarks introducing the second reading of the ban in the House of Assembly refer multiple times to the need for social licence, namely to increase public trust in the government's regulatory processes (Government of Nova Scotia 2014b). Thus, although the scientific uncertainty identified in the Wheeler report provided Younger with some degree of cover to justify the ban, unlike Gallant, Younger relied much more on the public's assertion that the risks of hydraulic fracturing were intolerable to legitimize his decision.

The Nova Scotia case illustrates the challenges facing elites acting in an uncertain policy space. Although moratoria have the potential to dampen political conflict, as policy positions moratoria also provide implicit support for ongoing challenges from environmental advocates pushing for higher levels of precaution and stringency. The lack of definitive policy direction in a moratorium provides cover for politicians to avoid affirming specific policy goals and potentially to avoid blame, however, the growing public resistance in Nova Scotia from 2012 to 2014 illustrates that this lack of direction does not always play well to a mobilized public. Calling for a moratorium was a successful strategy for Brian Gallant in New Brunswick, but the actions of the Liberal government in Nova Scotia illustrate that maintaining a moratorium may be less politically attractive for incumbent governments. In effect, arguably Younger learned that in response to a highly salient policy problem, the government needed to take political action, even if the response was disproportionate to the perceived policy risk among bureaucrats. Finally, the case confirms that decisions to implement bans are not necessarily prompted by processes of social policy learning that change politicians' causal beliefs regarding risk, instead, precautionary regulations can result from processes of political learning that change politicians' beliefs about what is politically feasible. Whether the subsequent policy reform is susceptible to erosion because of this process of political learning among elites remains an open research question for empirical testing and debate.

Chapter Eight

Regulating Uncertainty

Energy policy making in Canada has historically been that of quiet politics, influenced primarily by market conditions and determined "under the radar" by regulators, public servants, and government officials (Culpepper 2011; Gattinger 2012; Doern 2005; Hessing, Howlett, and Summerville 2005). There is no doubt that the contribution of the oil and gas industry to the Canadian economy and the potential for ongoing job creation weigh heavily in determining policy makers' interests (Gattinger 2015; Gattinger and Aguirre 2016; Graham, Daub, and Carroll 2017). At the same time, the energy sector is increasingly linked with environmental concerns and contentious politics (Lowry 2008; Neville 2021). Environmental mobilization based on cultural environmental world views (Douglas and Wildavsky 1983) and established activist networks puts significant pressure on policy makers to consider social, environmental, and health concerns prior to pursuing development (Gattinger 2015; Neville et al. 2017; Montpetit, Lachapelle, and Harvey 2016).

Hydraulic fracturing encapsulates these dynamics. On the one hand, hydraulic fracturing is a method of oil and gas extraction. As such the practice is simply a new technique that contributes to Canada's conventional oil and gas production. On the other hand, hydraulic fracturing also poses a variety of uncertain environmental and health risks (Council of Canadian Academies 2014; Atherton et al. 2014; LaPierre 2012; Cleary 2012; New Brunswick Commission on Hydraulic Fracturing 2016; Jackson et al. 2014). These elements reflect a classic tension of environmental policy making in which policy makers are faced with a series of trade-offs between acquiring economic benefits and mitigating environmental harms. What is new in the case of hydraulic fracturing is the additional element of uncertainty vis à vis technological risk (Gattinger and Aguirre 2016; Neville et al. 2017). As with other emergent technologies, hydraulic fracturing can generate substantial uncertainties concerning the assessment of economic and environmental costs and benefits. The speed of technological innovation, the volatility of global gas markets, and the paucity of environmental data have generated substantial variability with regard to the size of projected resources, potential jobs, and cumulative environmental effects (Small et al. 2014; Neville et al. 2017; Carter, Fraser, and Zalik 2017). Provincial governments have responded to this challenge in substantially different ways, resulting in significantly different regulatory frameworks across the country.

This book argues that a crucial element of how governments come to determine their priorities is their collective perceptions of risk. Differences in the social construction of risk trigger different processes of learning among policy makers, leading to different regulatory outcomes. Risk narratives include ideas about the extent of harm (dread risk) and the probability of occurrence, including the capacity to estimate said harms and probabilities (unknown risk) (Sunstein 2009; 2005; Klinke and Renn 2012; Blyth 2009; 2007). This study puts forward a typology of four different risk narratives generated by unknown and dread risk: *linear, complex, uncertain,* and *catastrophic*. Under conditions of *linear risk*, harm is minimal, and the probability of occurrence is low. *Complex risk* narratives are those in which probabilities are difficult to determine but can be estimated through scientific exploration and analysis. *Uncertain risk* narratives refer to those in which the severity of harms or the probability of occurrence is difficult to ascertain – risks that are on the edge of "unknown unknowns." Finally in *catastrophic risk* narratives, harm is severe and certain to occur.

This book traces the political economy of these "risk narratives," identifying some of the conditions under which these ideas emerge, charting how they filter through institutional structures and trigger alternate processes of learning among policy makers. To explore these contingent clusters the study examined four cases of provincial regulatory development in hydraulic fracturing: British Columbia, Alberta, New Brunswick, and Nova Scotia, from 2006–2016. Table 8.1 summarizes case study results.

The balance of industry power versus environmental mobilization helps to explain broad regulatory divisions between Canadian regions, determining variation between British Columbia's pro-development regulations on the one hand and moratoria and bans enacted in the Maritimes. Indeed, British Columbia, a province with a history of oil and gas development, has been content to layer new regulations onto the existing regulatory framework rather than devote resources toward developing completely new regulatory structures. Yet the relative balance of industry

Table 8.1. Summary of case study findings

Case	Industry Strength	Env. mobilization	Risk narrative	Institutional conditions	Causal mechanism	Regulatory outcome
BC 2006 –16	High	Low	Linear	Closed	Technical learning	Single-issue
AB 2006 –16	Medium	Medium	Linear/ Catastrophic	Closed/ Open	Technical / Political learning	Single-issue plus
NB 2010 –2013	Medium	Medium	Complex	Closed	Epistemic learning	Comprehensive regulation
NB 2013 –2014	Medium	High	Uncertain	Open	Social learning	Moratorium
NS 2010 –2012	Low	Medium	Uncertain	Closed	Technical/ Political learning	Moratorium
NS 2013 –2014	Low	High	Catastrophic	Open	Political learning	Ban

and environmental mobilization in each province creates the context – and an opportunity – for particular risk narratives to emerge. The promise of economic benefits can be countered by attention to environmental harms, especially as the salience of dread risks increases among the public, evident in Nova Scotia's precautionary approach from the get-go. It is the interaction of material economic interests with narratives about uncertainty that helps explain why policy makers prioritize environmental harms differently, especially across time. Ideas about risk empower actors to privilege certain regulatory designs over others.

In British Columbia the rapid and substantial interest of oil and gas players in the Montney and Horn River shales in the late 2000s, in the face of minimal environmental opposition, facilitated a narrative of linear risk among regulators within the BC Oil and Gas Commission. In Alberta, the nascent interests of the shale gas industry as compared with the complete dominance of the oil sands, together with attention to dread risks in other jurisdictions provided an opportunity for regulators to experiment with more comprehensive regulatory options. In New Brunswick in 2013, the substantial stake of SWN in mapping the resource, together with an unease among bureaucrats regarding potential environmental risks, led to the initial social construction of hydraulic fracturing as a complex risk. In Nova Scotia in 2012 the lack of industry engagement, together with rapidly increasing salience of dread risks among activated publics fostered narratives of uncertain risk among bureaucrats in the Department of Energy and the Department of the

Environment, as well as among the governing party leaders. Similarly in New Brunswick in 2014, even though industry interest remained strong, the high salience of environmental concerns encouraged government officials in opposition to put forward a narrative of scientific uncertainty.

Risk narratives set actors down a particular path, opening some regulatory responses while closing others, pathways which were facilitated and reinforced by institutional conditions. Regulators in both British Columbia and New Brunswick were set on a course to seek out regulatory solutions to the problem of encouraging development while mitigating potential environmental harms. In both cases, bureaucrats were insulated in the initial stages from broader political debates, functioning within the industry regulator or the ad hoc working group. This insulation provided civil servants with the time and resources to draw on their own scientific expertise and the experiences of regulators, or to consult with the broader epistemic community to develop new policy designs. The dominant risk narratives in each case guided officials' scope of learning. In British Columbia linear risk narratives constrained bureaucrats to focus on instruments and settings (e.g., well construction spacing), while in New Brunswick, complex risk narratives emboldened the members of the Natural Gas Group and other officials to consider a wider range of risks (e.g., public health). In British Columbia where learning was technical, single-issue regulation was adopted. In New Brunswick the broader scope of epistemic learning resulted in a more comprehensive regulatory approach.

Conversely, once policy processes were opened to the public through the ad hoc process of the New Brunswick 2012 "Listening Tour," or the external review in Nova Scotia in 2013, risk narratives helped actors, in this case leaders of opposition, environmental advocates, and Indigenous leaders, to construct and generate political support. The perception of scientific uncertainty, together with the potential for catastrophic environmental harms, provided grist for anti-fracking activists to call into question the credibility of both science and the state. Throughout consultations in New Brunswick in 2012 and town hall meetings in Nova Scotia in 2013, residents, academics, and environmental advocates drew on their own scientific studies to challenge the perception that the risks of hydraulic fracturing could be contained or managed through regulation, and by extension that the government could manage those risks. Alberta provides an uneasy in-between case, with regulators engaging in processes of technical learning to address perceived linear risks, but also responding to industry's attention to public concern, prompting more political learning regarding stakeholder engagement around cumulative effects.

Figure 8.1. Risk and hydraulic fracturing regulation in Canada.

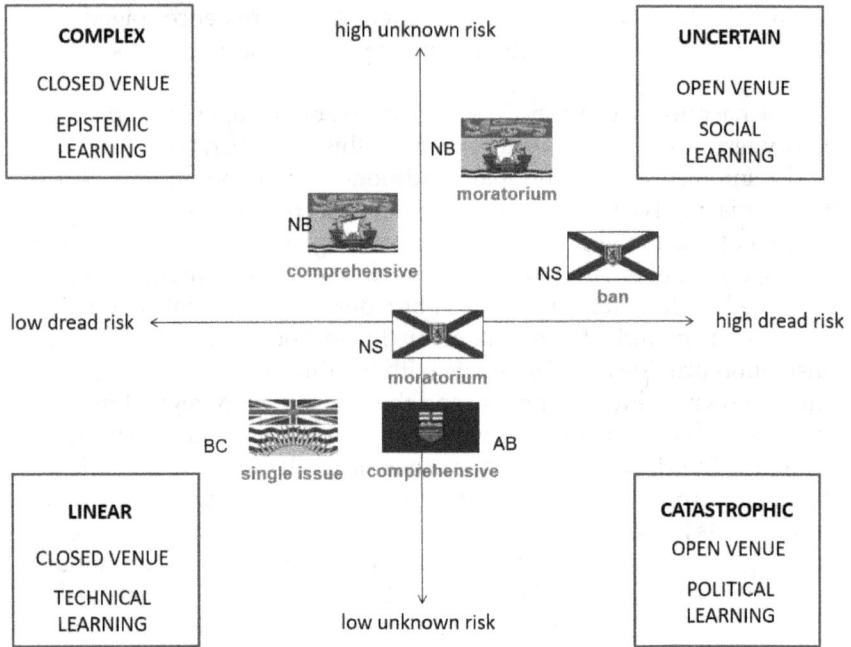

In the absence of tools to resolve thorny "unknown unknowns," moratoria and bans seemed like more politically feasible options for incumbent policy makers trying to maintain power. In New Brunswick, instituting a moratorium provided political cover for Brian Gallant, who was concerned about scientific uncertainties as well as the political risks of exacerbating public opinion, as is evident in the five conditions set for lifting the moratorium. In Nova Scotia, Andrew Younger's speedy implementation of a legislated ban in 2014 reflected concern about the political risks of engaging in hydraulic fracturing, resulting in a more stringent regulatory approach than we would otherwise anticipate.

The cases illustrate two key dimensions of the provincial regulation of hydraulic fracturing in Canada: first, the ways that industry and environmental interests, institutional conditions, and risk narratives cluster together to foster the development of regulatory frameworks, and second, the processes of learning through which these frameworks are formulated. Figure 8.1 plots these processes and instances of regulatory change.

In the cases where the policy-making process was relatively insulated, regulators engaged in processes of technical and epistemic learning.

What distinguished the experiences of government officials in British Columbia, Alberta, and New Brunswick was the comprehensiveness of the search. I find that the process of learning in each case was influenced by risk perception. In British Columbia, the environmental risks were perceived to be minimal, a tolerable problem to gain economic benefits. In Alberta, the risk of seismic harms and well-to-well contamination were understood to be emerging and potentially scientifically uncertain. In New Brunswick the uncertainty regarding environmental harm led to a more extensive and wide-ranging process of epistemic learning.

Conversely, in cases where regulatory design was more open because of ad hoc consultations, working groups, and external review processes, policy makers engaged more fully in processes of social and political learning. Here again processes of learning were influenced by risk narratives. In Alberta the contingent restructuring of the ERCB together with rapidly emerging catastrophic risk narratives in other jurisdictions allowed regulators to adopt more wide-ranging regulatory frameworks than in British Columbia. For the Dexter government in Nova Scotia, a combination of dread and unknown risk facilitated the decision to kick the problem down the curb to the next government. Without scientific direction respecting the most efficient solution, the government decided to hit the pause button, instituting the de facto moratorium. Similarly for the Gallant government, scientific uncertainty justified the pursuit of a moratorium, effectively implementing an interim solution to assuage the public's fears of dread risks while asserting that future development might be possible. Alternatively, in the case of the Liberal government in Nova Scotia, the pitch of resistance generated by environmental mobilization resulted in the dominant narrative of catastrophic risk; that is, that probable harms were severe and certain to occur. Once that narrative gained traction among the public, the policy options available to Minister Younger were limited: in effect, either to maintain the moratorium or to implement a legislated ban. The ultimate choice to implement a ban demonstrated a symbolic action that the government was responding to the public's fears.

Regulation and Risk Narratives in Other Jurisdictions

The relationship between risk narratives, processes of learning, and regulatory reform traced in this book was developed to explain provincial variation in hydraulic fracturing regulation in the four provinces under study. Yet risk narratives have the potential to explain politics of fracking in other jurisdictions, both within Canada and internationally. The following vignettes demonstrate that when different risk narratives emerge across jurisdictions, certain pathways are more or less likely, resulting in differences in the extent of policy change.

Saskatchewan

As discussed in chapter 3, Saskatchewan has engaged in very little regulatory reform or policy change, despite the unprecedented boom in oil production from hydraulic fracturing in the Bakken oil fields in the 2000s. Statements by elected officials reflect a narrative of linear risk, with provincial ministers reiterating frames regarding the "50-year history of safe use" of the practice and the efficacy of existing oil and gas regulation (Campbell, quoted in Carter and Eaton 2016, 403). As in British Columbia, elected officials in Saskatchewan have also consistently reiterated the economic benefits of the oil and gas industry for the province (Olive and Delshad 2017; Olive 2016; 2018). In Saskatchewan this support has been bipartisan, with little opposition within the provincial legislature to development from any parties (Carter and Eaton 2016). A strong anti-fracking movement, coalesced around either environmental risks or scientific uncertainty, has failed to emerge in the province, partially due to the focus of environmental organizations on other issues and partly due to the limited mobilization of rural landowners against the practice (Carter and Eaton 2016; Eaton and Kinchy 2016; Olive and Valentine 2018). As in British Columbia, the early dominance of a linear risk narrative among industry representatives and elected officials has constrained the ability of environmental advocates to generate a compelling narrative of uncertain or catastrophic risk needed to support a groundswell of public attention to local environmental harms.

Quebec

Quebec stands out as the first province to declare a moratorium on hydraulic fracturing, which it implemented in 2011 and reaffirmed in 2014 and 2018 respectively (Montpetit, Lachapelle, and Harvey 2016; Kestler-D'Amours 2018). The political dynamics reflect similarities to the Maritime cases in that despite (or because of) Quebec's limited experience with conventional oil and gas development, elected officials were initially bullish on the potential economic benefits of hydraulic fracturing. Nevertheless, as in Nova Scotia, public hearings through the public environmental agency (the Bureau d'audiences publiques sur l'environnement, or BAPE) served as a catalyst for environmental activism. Anti-fracking groups dominated the hearings, and were "twice as likely to be represented in official government public consultations"

(Montpetit, Lachapelle, and Harvey 2016, 63). Prominent anti-fracking arguments centred on claims that "hydraulic fracturing was too uncertain to authorize shale gas development," reflecting an uncertain risk narrative, and "hydraulic fracturing was likely to contaminate groundwater," reflecting a catastrophic risk narrative (Montpetit, Lachapelle, and Harvey 2016, 70).

In its report on the outcomes of the hearings, the BAPE reiterated the uncertain risk narrative that existing scientific knowledge was "insufficient" and that provincial capacity to regulate the issue was lacking. Following the release of the BAPE report, the provincial government implemented the moratorium in the summer of 2011 (Montpetit, Lachapelle, and Harvey 2016). As in Nova Scotia, Quebec-based anti-fracking advocates used an institutional forum – here the public hearings of the BAPE – to highlight the scientific uncertainty of the technology and to stress the potential for risks to environmental health (Chailleux 2020). The proliferation of these narratives was further supported by an opening up of the institutional environment, as subsequent BAPE hearings expanded the scope of inquiry to include a Strategic Environmental Assessment and a second assessment examining the overall viability of the shale gas industry, which included studies of social movements against fracking within Quebec (Chailleux 2020, 10). Ultimately the groundswell of public opinion against the practice prompted elected officials to implement and consistently reaffirm the moratoria to diminish the potential for electoral loss, reflecting a mode of political learning (Chailleux 2020; Montpetit, Lachapelle, and Harvey 2016).

Newfoundland and Labrador

The relationship between catastrophic risks and precautionary regulation is also evident in the province of Newfoundland and Labrador. In December 2013 the government of Newfoundland and Labrador implemented a moratorium on hydraulic fracturing, an action which was surprising considering the significance of Newfoundland and Labrador's extensive offshore oil and gas industry to provincial policy makers and the provincial economy (Carter 2020). Public attention to the issue was initially sparked by an application for an exploration licence. A Canadian petroleum company, Shoal Point Energy, and its partner, Black Spruce Exploration, filed an application to use hydraulic fracturing techniques in onshore-to-offshore wells in Western Newfoundland (Carter and Fusco 2017). In response, environmental activists and local groups began organizing around groundwater risks, building an anti-fracking campaign focused on dread risks regarding contamination of drinking

water, as well as concerns around harms to fishing and tourism industries in and surrounding Gros Morne, a UNESCO World Heritage Site
(Carter and Fusco 2017, 112–14). Public hearings provided a venue for
catastrophic risk narratives to proliferate, which were bolstered by perceptions that provincial staff were ill- equipped to consider the range of
environmental uncertainties generated by hydraulic fracturing (Carter
and Fusco 2017). The rapid spread of catastrophic risk narratives served
to generate unprecedented levels of public mobilization against fracking, surprising provincial government officials who had anticipated
a simple environmental assessment process (confidential interviewee
2014j). Ultimately, the decision to implement a moratorium was influenced by processes of political learning among regulators and politicians (Carter and Fusco 2017). Although the process of political learning
did not lead to a legislated ban, a subsequent independent review in
2016 recommended that the government keep the moratorium in place
indefinitely, resulting in a de facto ban (Government of Newfoundland
and Labrador 2016; L. Bird 2016).

United Kingdom

Beyond Canada, there are aspects of the framework that are also appli-
cable to the political dynamics of hydraulic fracturing in the United
Kingdom. The position of the national government, which regulates
shale gas development in England, has varied considerably over the last
decade, from implementing an initial moratorium in 2011, to aggres-
sively promoting shale gas development from 2012 to 2018, to adopting a renewed moratorium in 2019 and 2022 (Bradshaw et al. 2022;
Evensen 2018; Williams, Martin, and Stirling 2022).[1] England's conventional onshore oil and gas industry has historically been unable to keep
pace with domestic energy demands, with onshore wells peaking at 162
wells drilled in 1943, but gradually declining from 1985 onwards (Hayhurst 2023; Bradshaw et al. 2022). In 2010 the government issued an
exploratory licence to Cuadrilla Resources, an Australian-owned company based in the United Kingdom to investigate resource potential in
Lancashire (Evensen 2018; Cudrilla Resources n.d.). However, in 2011,
initial test drilling resulted in seismic activity near Blackpool, which led
the UK government at the time to implement a temporary moratorium
while it investigated the potential environmental risks of the practice.
Subsequent internal and external reports, namely from the Royal Society and the Royal Academy of Engineers concluded that the risks from
hydraulic fracturing were tolerable, leading to the lifting of the moratorium in late 2012 (Bradshaw et al. 2022). Similar to governments in

British Columbia, the UK government subsequently adopted a bullish pro-development stance, with Prime Minister David Cameron firmly declaring that the government would be "going all out for shale" in 2013 (Bradshaw et al. 2022; Whitton et al. 2018). Studies on framing and coalition dynamics in the United Kingdom have traced the dominance of economic benefits, energy security, "manageable risk" (e.g., "fracking is safe"), and clean energy frames used by UK elected officials and industry representatives (Bomberg 2017a; 2017b; Williams and Sovacool 2019; Jaspal and Nerlich 2014), suggesting the use of linear risk narratives very much in line with the talking points put forward by BC elected officials. In their study of public participation processes regarding hydraulic fracturing, Williams et al. (2022) identify the dominance of instrumental decision making within the UK government from 2012 to 2019 in which "processes typically allowed only a narrow range of issues to be considered and there was limited scope for public influence on shale development policy" (11). These findings reflect a type of technical learning among regulators in the United Kingdom that is consistent with the linear risk narrative put forward in this book.

In contrast to British Columbia however, the linear risk narrative put forward by UK officials failed to entrench among the electorate, leaving the door open for alternative risk narratives to take hold. Anti-fracking coalition members, including environmental advocates, local campaign groups in Lancashire, as well as elected governmental officials of opposition parties (including both Labour and the UK Green Party), have drawn on both catastrophic and uncertain risk narratives to challenge the pro-development stance of the UK government. Frames put forward by anti-fracking campaigners included environmental harms such as groundwater pollution, seismicity, and land contamination. In addition to these catastrophic harms, an "elusive threats" frame also emerged regarding the uncertain risk of the new technology and concerns regarding the capacity of governments to regulate the practice (Williams and Sovacool 2019; Jaspal and Nerlich 2014; Bomberg 2017a). Despite the initial narrow scope of government consultation, these frames proliferated through public debate, heightening to a fever pitch when Cuadrilla's first commercial drilling project at New Preston Road in Lancashire triggered over 134 seismic events in August 2018 (Bradshaw et al. 2022). Catastrophic narratives regarding earthquakes paired with the ongoing frustrations of UK residents with local planning and policy processes, which served to further undermine trust in the national government. Demonstrating a process of swift political learning, the UK Conservative party ultimately implemented a moratorium prior to the general election in 2019, citing "scientific uncertainty over the probability and magnitude of further seismicity" as the key rationale for the regulatory change (Williams, Martin, and Stirling 2022, 2).

United States

As discussed in chapter 2, US states have implemented a wide range of regulatory responses to hydraulic fracturing, with significant differences between the relatively lax regulatory practices of energy dominant states of Texas and Wyoming, and states that established early moratoria on the practice following vehement public debates, such as New York State (Rabe 2014). In her comparison of framing debates in the United States and the European Union, Bomberg (2017b) highlights tendency of the majority of elected state officials, in both legislative and executive roles, to adopt a pro-development stance, citing both economic benefits and minimal risk narratives in support of their positions. Conversely, anti-fracking activists in the United States were more likely to draw on catastrophic narratives of water contamination as well as anger towards the oil and gas industry in public debates (Bomberg 2017b; Pierce et al. 2022).

The state of Colorado has had one of the more complex cases of regulatory development in the United States. Despite a significant production boom in the late 2000s, during the early 2010s the Colorado Oil and Gas Conservation Commission (COGCC) implemented a broad array of regulatory reforms, including chemical disclosure, well setbacks, and baseline water testing regulations (Weible and Heikkila 2016; Heikkila et al. 2014), reflecting a form of comprehensive regulation. Nevertheless, in 2019 the state legislature implemented a sweeping reform of the COGCC, formally changing the mission of the organization to one of protecting public health and the environment (Pierce et al. 2022). Policy narrative scholars have drawn on a variety of concepts to explain these dynamics, including policy core beliefs, heroes and victims, and emotions (Weible and Heikkila 2016; Pierce et al. 2022), however a closer attention to risk narratives, institutional venues, and learning provides deeper insight into the case.

Comparative research into the Colorado case demonstrates that unlike their counterparts in energy dominant states such as Texas, elected officials in Colorado were more likely to acknowledge the potential for environmental risks deriving from hydraulic fracturing, including well-to-well contamination, flaring, and land use impacts, suggesting a complex risk narrative, similar to early stages in New Brunswick (Weible and Heikkila 2016). In particular, the Democratic state governor, John Hickenlooper, had prior industry experience and training as a geologist and was well aware of the potential for political risk from public unease with fracking (Heikkila et al. 2014). Hickenlooper strongly supported informal stakeholder engagement within the COGCC between the oil and gas industry and environmental groups in order to shape regulatory reform, including new rule making for chemical disclosure, groundwater monitoring, increased setbacks, and substantial methane policy reforms (Heikkila et al. 2014; Rabe 2022).

At the same time, environmental and citizen groups in Colorado mobilized regarding catastrophic risk narratives, lobbying at both local and state level venues to expand the conflict beyond the COGCC (Pierce et al. 2022; Bomberg 2017b). Environmental activists eventually succeeded by gaining the support of two senior Democrats in the Colorado state legislature to introduce substantial institutional reforms to the COGCC, holding four public hearings in 2019 (Pierce et al. 2022). In their analysis of the public hearings, Pierce et al. (2022) find that the anti-fracking coalition predominantly used catastrophic risk narratives regarding environmental health impacts in their testimony, reflecting a strong combination of anger and fear targeting the oil and gas industry. Ultimately, the anti-fracking coalition persuaded legislators to pass the bill, significantly changing the COGCC's mission from economic development to that of "protect[ing] public health, safety, welfare, the environment and wildlife resources" (COGCC, in Pierce et al. 2022). The Colorado case demonstrates a blend of both epistemic and political learning among bureaucrats and officials, leading to more expansive regulatory reform over time.

These sketches illustrate the portability of the framework to other jurisdictions, drawing attention to the patterned interaction of risk narratives, institutional conditions, and policy change. Policy makers' attempts to contain regulatory reform within bureaucracies or arm's-length commissions often draw on narratives of linear risk, with elected officials stressing frames of manageable risk along with proposed economic benefits. Different institutional structures can foster insularity. Although an independent regulator can insulate rule-making processes from the vagaries of political debate, governments can also establish ad hoc teams to insulate bureaucrats focused on instrumental policy learning. Conversely, several of the cases illustrate the ways in which catastrophic and uncertain risk narratives can increase the salience of environmental harms in the public eye and bolster the legitimacy of anti-fracking activism. When coupled with a catastrophic risk narrative, even a temporary venue such as an external review, a legislative hearing, or a local planning application can generate sufficient political heat that policy makers will engage in extensive regulatory reform.

These findings provide nuance to foundational policy process literature on the role of institutional conditions and venues in expanding and or mitigating opportunities for policy change (Schattschneider 1960; Baumgartner and Jones 1993; Pralle 2006a; Stephan 2020). Although scholarship has tended to focus on the venues generated by federalism or the courts, the cases studied here illustrate the analytical value in revisiting the interaction of framing and narratives with administrative features such as the division of responsibility across departments, ad hoc working groups, public consultations, and independent agencies (Hoberg and Phillips 2011). More research is needed into the ideational factors spurring policy makers' decisions to broaden decision-making processes, as well as the unintended effects of administrative features on subsequent policy design (Pralle 2006a; Janzwood 2021). Of import is also the role of local debates in building a groundswell of support behind particular catastrophic narratives, which, despite their symbolic nature can take hold within subnational or national venues with greater regulatory authority (Neville 2021; Bomberg 2017a; Evensen 2018).

The findings of the main case studies of the book and the vignettes outlined above suggest the need for more nuanced consideration of the interplay of industry power, environmental mobilization, and the emergence of risk narratives, particularly with regard to sequencing and pace (Pierson 2004). Although energy dominant states are inclined to develop linear risk narratives, the Alberta and Colorado cases demonstrates that under some conditions, environmental risks and policy failures in other jurisdictions can prompt regulatory reform, even in the face of substantial industry power. For jurisdictions in which industry strength is less consistent – whether due to volatility in the global price of gas, changes in geopolitics, uncertainty in resource estimates, or a tentative financial climate – pro-development policy makers may find it difficult to sustain and entrench a narrative of linear risk over time. As the Maritime, Newfoundland, and UK cases illustrate, elected officials' pro-development stances are much more tenuous in states with a limited history of oil and gas development than classic analyses of business power might suggest (Lindblom 1980). Under these mutable conditions, it may be that the sequencing or layering of economic benefits and environmental harms is more important than the relative power of industry or environmental mobilization in determining when and why risk narratives emerge. Future research might tease out the causal mechanisms that shape the emergence of risk narratives, particularly regarding the connections between uncertainty, urgency, and the resources of competing coalitions and interest groups (Lesch and Millar 2022; Lesch 2021; Chailleux 2020; Moyson et al. 2022).[2]

Risk Narratives, Policy Studies, and Environmental Politics

The focus of this study on risk narratives contributes to a burgeoning area of policy studies that move "beyond the literature's initial focus on the notion that 'ideas matter,' to explorations of how they matter – and, in particular, how they intersect with power relations" (Parsons 2016, 446; Béland, Carstensen, and Seabrooke 2016; Blyth 2016). Much has been written in political science regarding the role of epistemic uncertainty in guiding how policy decision makers come to understand their interests (Hall 1993; Blyth 2006; Haas 1992; 2004; Dunlop 2017). Yet often in policy studies, and particularly in policy analysis, uncertainty is considered to be primarily an objective condition, determined and measured through scientific analysis, instead of a socially constructed phenomenon (Wellstead, Cairney, and Oliver 2018). This study bridges this gap by parsing out the different elements of uncertainty and theorizing how ideas about unknown and dread risk shape how political actors come to understand their interests. Although the study describes the political and economic conditions under which these different narratives emerge in Canadian provinces, the primary focus of the analysis is on how risk narratives increase the legitimacy of different sets of actors, shape pathways of policy formulation and learning, and generate stringent (or more lax) regulatory reforms.

Risk Narratives, Legitimacy, and Political Actors

Risk narratives generate political power for some sets of actors, diminishing the clout of others in the policy process. As such, this study demonstrates how the content of policy ideas amplifies or hinders the resources of different sets of actors within the policy-making process (Béland, Carstensen, and Seabrooke 2016). My research suggests that risk narratives can strengthen the influence of different actors by bolstering their perceived credibility or legitimacy within the policy process.

One of the persistent dynamics across the cases is that narratives of complex risk are often complemented with a turn toward expert authority, increasing the influence of epistemic communities (e.g., hydrologists, seismologists, geologists) in the policy subsystem. Conversely, uncertain and catastrophic risk narratives often correlate with a lack of deference to expert authority and a corresponding turn to popular authority (Skogstad 2003). The cases of New Brunswick and Nova Scotia, and to some extent Alberta, illustrate that catastrophic risks are also often contested risks, as actors question the credibility of expert authority to provide appropriate policy solutions. In New Brunswick, the Alward government's reliance on scientific expertise to resolve public conflicts made the government particularly vulnerable to challenges rooted in dread risk. This evidence calls into question the assertion made by some risk scholars (Sunstein 2005; 2009) that in the face of heightened public fears regarding catastrophic harms, government leaders should turn to experts to determine policy making processes. Although putting scientists in the driver's seat is likely effective from a policy perspective, for any government leader wanting to get re-elected, depending on scientific certainty to resolve political uncertainty is a tenuous proposition at the very best. If the public perceives that the likelihood of a severe harm is certain and the government has failed to act appropriately, trust in government intentions – and institutions – can quickly erode (Neville and Weinthal 2016; Neville 2021). Faced with these conditions, policy makers are incentivized to implement more and more stringent policies to gain back the public's trust. These findings suggest that there is a strong relationship between different risk narratives and deference to expert authority (or lack thereof).

The study also finds that risk narratives can bolster the popular authority of other actors in the policy subsystem. In New Brunswick and Nova Scotia, narratives of uncertain and catastrophic risk were used by environmental advocates to justify increased public engagement in policy decision making, strengthening their arguments that the policy decision-making process needed to be more open, reflexive, and democratic. More research is needed regarding the mechanisms through which risk narratives influence the legitimacy of different actors in the policy subsystem and how increases in authority can shape actors' input on policy decision making and design (Schmidt 2013; Doberstein and Millar 2014). Further research could also elucidate successful models of public consultation with highly activated and mobilized publics (Stephan 2020; 2017; Williams, Martin, and Stirling 2022; Fast 2016a).

One of the limitations of this study is that the focus on policy makers and regulators elides some of the complexities of the anti-fracking movement, particularly regarding the opposition of Indigenous nations. In New Brunswick, the Elsipotog protests initially centred on narratives of catastrophic risk regarding hydraulic fracturing, but rapidly grew to incorporate concerns regarding sovereignty and failures of reconciliation. Indigenous mobilization against fracking can draw on popular authority, but it also draws on political authority rooted in Indigenous sovereignty, treaties, and Indigenous rights and title (Garvie and Shaw 2014; Garvie, Lowe, and Shaw 2014). When do provincial policy makers incorporate Indigenous resistance into their political learning calculations? For example, do risk narratives that incorporate legal risks –

for example, those posed by the duty to consult – prompt more urgent action among policy makers?[3]

Beyond experts and interest groups, this study highlights the role journalists play in the process in amplifying risk narratives. By and large, the study relies on the analytical assumption that news media content functions as a proxy for public opinion in the eyes of elites, driving attention to a particular policy problem or issue frame (Baumgartner and Jones 1993; Pralle 2003; Baumgartner, Boef, and Boydstun 2008; Pralle and Boscarino 2011; Wolfe 2012; Olive and Delshad 2017). Indeed, the volume of coverage in New Brunswick and Nova Scotia on hydraulic fracturing lends credence to the common assumption among policy scholars that quantity of media attention can drive up issue salience following a focusing event (Birkland 1998; Kingdon 1995; Wolfe 2012). At the same time, the BC case illustrates that media can also have a stabilizing effect on policy subsystems, by locking-in particular narratives and minimizing others (Wolfe 2012; Montpetit, Lachapelle, and Harvey 2016). In their analysis of advocacy coalitions influencing hydraulic fracturing policy making in British Columbia and Quebec, Montpetit et al. (2016) argue that journalists have a privileged position in selecting particular frames that generate attention in some cases (Quebec) while dampening issue salience in others (British Columbia). Rather than simply relaying frames generated by experts or environmental advocates, they suggest that journalists rely on their own professional codes of objectivity and other institutional and professional norms to identify what makes a "good story." In aggregate, these decisions can drive overall media attention. My findings from the Nova Scotia case provide support for this proposition, as a significant proportion of arguments regarding the risks of hydraulic fracturing were attributed to journalists reporting on other jurisdictions, rather than to local experts or interest groups. Conversely, media coverage in New Brunswick was much more attentive to a range of claims from a variety of provincial and local actors. I argue that this finding reflects the intensity of public opinion and debate that occurred locally within New Brunswick. These differences between the Nova Scotia and New Brunswick cases suggest that although journalists may have a privileged position in setting the agenda, they do not always determine public debate (Renn 2008).

More importantly, I concur with Montpetit et al.'s observation that "it is more the characteristics of issues, rather than subsystem actors, beside journalists themselves, which drive media coverage" (Montpetit, Lachapelle, and Harvey 2016, 78). As noted above, my reading of the Quebec case is that narratives of uncertain and catastrophic risk facilitated processes of political learning among politicians, resulting in the moratorium. These findings suggest that policy process scholars would be wise to turn to a longstanding research tradition on the role of media in the social amplification of risk (R. E. Kasperson et al. 1988; Slovic 1987; 2000; J. X. Kasperson et al. 2003; Breakwell 2007; Sutton and Veil 2017). Risk scholars argue that professional news media act as amplification "stations" that can either amplify or attenuate public attention to risk. News media can amplify risk perception by increasing the volume of coverage, emphasizing contestation regarding the credibility of information, and dramatizing events (R.E. Kasperson et al. 1988). Research has found that professional media reports become more relevant to public opinion "the more that access to physical evidence or direct personal experience is lacking" (Renn 2008, 129). More recently, scholars have argued that social media act not only as an amplification station but also as a source of direct experience, as far flung publics can vicariously experience natural disasters, explosions, and violent conflict through personal videos taken on cell phones and uploaded online (Sutton and Veil 2017). Studies of dissemination of risk narratives through social networks in real time, for example by examining Twitter feeds, would help further identify the characteristics of risk narratives that are easily passed on versus those which are not (Sutton and Veil 2017).

Risk Narratives, Diffusion, and Learning

With regard to learning, the case studies support the presumption that institutional and ideational factors interact to predict different modes of learning, affirming recent theoretical advancements (Dunlop and Radaelli 2018a; 2018b; Trein 2018; Weible 2008; Weible and Nohrstedt 2012). The comparative structure of this study has focused primarily on regulatory reform within cases, focusing on the relationship between risk narratives and processes of learning. But one of the interesting findings of the case studies is the diffusion of both technical expertise and catastrophic risk narratives across jurisdictional boundaries, with regulators in Canada attending to developments in the United States (Millar 2021), UK activists drawing on US policy failures (Jaspal and Nerlich 2014; Williams and Sovacool 2019), and provincial politicians drawing on the experiences of their ideological peers, rather than their geographic neighbours. One of the more unexpected findings in the Maritime cases is that despite the contextual similarities between the two provinces, partisan politics mattered with regard to regulatory outcomes in New Brunswick, but not in Nova Scotia. These findings suggest that partisanship, or at least general policy core beliefs, are less predictive of regulatory outcomes than specific risk narratives. Nevertheless, similarities in the party affiliations of both Brian Gallant and Andrew Younger suggest the possibility that provincial policy makers may look to their ideological peers when assessing what is politically feasible. More research is needed concerning the dynamics of interjurisdictional information sharing, particularly with regard to the political learning and heuristics of representativeness and availability and the influence of perceived geographic and/or ideological closeness in Canada (Weyland 2005; Boyd 2017; Boyd and Olive 2021; Lesch and Millar 2022).

This study also demonstrates that contrary to linear models of knowledge mobilization, policy makers do not necessarily have to believe in deep epistemic, causal uncertainty to engage in radical policy change. Instead, radical policy change often stems from processes of political learning. Once electoral or political uncertainties arise, they often override policy makers' causal beliefs about the efficacy of a given policy, evident in the cases of New Brunswick and Nova Scotia. The study findings illustrate the limits of epistemic learning in triggering radical policy change. Instead, the research finds that policy makers' perceptions of political feasibility are ultimately a better barometer than perceptions of policy efficacy in predicting whether policy makers are likely to implement a stringent policy design. Future research might tease out how different sequences of learning generate regulatory change. For example, the case of Nova Scotia illustrates that processes of social learning can be tenuous and difficult for policy makers to manage. In situations where the perceived policy problem is uncertain, and in which the authority of government to address the problem is challenged by either expert or public authority, it can be difficult for policy makers to maintain an open-ended process of policy deliberation, leading instead to more strategic processes of political learning or bargaining. Whether modes of social learning always sow the seeds for political learning is a matter for future empirical inquiry.

The findings here also suggest the need for more refined models of the individual in policy learning (Moyson, Scholten, and Weible 2017; Millar, Lesch, and White 2019). This study's framework assumes that policy makers "are intendedly rational" and goal following, with the capacity to determine their material and electoral interests; but that they are susceptible to environmental pressures that affect issue salience, attention, and time (Jones 1999). I argue that risk narratives have a specific effect on issue salience, allocating elites' attention to some dimensions of a particular policy problem while eliding others. At the same time, risk narratives can activate publics, fostering authority contests which place pressure on policy elites to act in a decisive and timely manner. This analytical approach assumes that external factors have a constraining and shaping influence on elites' cognitive processes of decision making as expressed through policy formulation and design. Fundamentally, discussions of learning turn on the assumption that actors are updating – however inefficiently – their beliefs. Yet findings from cognitive psychology suggest that in addition to external factors, policy elites may also be influenced by unconscious, or "irrational" factors such as heuristics, emotion, deep core beliefs, and rules of thumb (Weyland 2005; Cairney and Weible 2017; Jones 2017; Moyson 2018; Moyson, Scholten, and Weible 2017; Taber 2003; Wilson 2011; Kahneman 2013; Tversky and Kahneman 1981).

Despite a widespread acknowledgement that these factors have a deep and enduring influence, the mechanism by which these factors influence elite decision making, at least within policy process theories, is unclear. For example, there is a rich tradition in political communication literature on source and framing effects (Druckman 2001a; 2001b; Druckman and Lupia 2000), but this has only recently begun to be integrated into Canadian public administration and policy studies (Lachapelle, Montpetit, and Gauvin 2014; Gravelle and Lachapelle 2015; Doberstein 2017). One avenue forward is to examine the boundary between learning, in all its modes, and emulation (Boushey 2010). Despite an extensive research agenda on mechanisms of emulation in policy diffusion research (Dolowitz and Marsh 1996; 2002; Dobbin, Simmons, and Garrett 2007; Simmons, Dobbin, and Garrett 2008), studies have documented significant conceptual overlap and lack of measurement in how learning and emulation are operationalized (Gilardi 2010; Maggetti and Gilardi 2016). A promising solution would be to examine how urgency in individual cognition (Kahneman 2013) might underpin collective processes of policy emulation (Lesch and Millar 2022).

Implications for Policy Makers

These findings have practical implications for policy makers working in energy policy making in Canada, particularly respecting the design of public consultation processes around risk. The cases documented in this book firmly put to rest the tired notion often put forward by practitioners that additional scientific information will resolve policy disagreements. Although this argument has been long dispelled by science and technology scholars, risk communication researchers, and political scientists studying public consultation mechanisms (Jasanoff 2003; Pielke 2007; Renn 2008; Hartley and Skogstad 2005; Skogstad 2003), it still has legs, as evident in the comments of several interviewees that the public was simply "misinformed" with regard to the risks of hydraulic fracturing (Alward 2014; Leonard 2014; Northrup 2014; confidential interviewee 2014a; 2014d). The New Brunswick and Nova Scotia cases illustrate the many ways in which a variety of actors within policy subsystems drew upon scientific authority to bolster their claims; sometimes because the evidence supported individuals' deep core beliefs, but sometimes because it served a strategic purpose by enabling actors to undermine their opponents' credibility. Although tempting to assert that this is (and has always been) the cost of doing politics, social science research suggests that the intersection of scientific expertise and citizen participation through modes of bargaining has the potential to further erode public trust in government institutions (Neville and Weinthal 2016; Fast 2016b; Doberstein and Millar 2014; Neville 2021). In particular, the Nova Scotia case demonstrates how even when consultation processes are carefully designed (Wheeler et al. 2015) they can foster institutional opportunities for adversarial bargaining. Moreover, the findings here suggest, along with other energy studies, that there is a reciprocal relationship between risk frames and perceptions of trust or credibility (Lachapelle, Montpetit, and Gauvin 2014; Fisk 2013; Thorn 2018). As the New Brunswick case illustrates, historical experiences of procedural "unfairness," whether through failed environmental assessments or restructuring of public institutions, can serve to exacerbate citizens' risk perceptions, focusing their attention on catastrophic risks. In practical terms, my research suggests that policy makers dealing with catastrophic or uncertain risk narratives need to pay close attention to the procedural dimensions of consultation processes, especially in clarifying the mechanisms through which public engagement processes will (or will not) feed into policy formulation and decision making (Millar 2013; Doberstein and Millar 2014; Doberstein 2016; Schmidt and Wood 2019; Williams, Martin, and Stirling 2022).[4]

Canada's energy policy-making system is undergoing a substantial transformation in which policy makers are already facing significantly low levels of public trust (Cleland and Gattinger 2017; Cleland et al. 2016; Neville and Weinthal 2016). At the same time, provincial governments in Canada are under increasing pressures to respond to and engage with societal concerns, whether from Indigenous nations, municipal governments, or local resident groups (Carter and Eaton 2016; Carter, Fraser, and Zalik 2017). This raises normative and empirical questions as to whether processes of collective public puzzling about means and ends should and could be used to increase levels of trust among citizens (Millar, Davidson, and White 2020; Levac and Wiebe 2020). How might provincial policy makers use mechanisms of reflexive learning to build trust? Do processes of collective puzzling lead to more effective policy designs and outcomes? When does social learning exacerbate conflict? Although maintaining ambiguity as to the means and ends of policy design likely facilitates collective puzzling, uncertainty is a very difficult position for governments to maintain for fear of undermining the state's credibility to act decisively in the face of perceived risk. Further exploration of the dynamics of social learning, or collective puzzling, can serve to deepen our understanding of the link between policy design and public acceptance of new technologies (Paterson, Tobin, and VanDeveer 2022; Walker et al. 2023; Walker and Baxter 2017).

This study demonstrates that the design of provincial regulatory frameworks for hydraulic fracturing is the outcome of a process through which policy makers determine their interests. Faced with a combination of economic, social, environmental, and scientific uncertainties, policy makers engage in different processes of puzzling and powering to guide their decisions. Sometimes regulators rely on their past experiences or scientific experts, but sometimes they turn to the political experiences of regulators in other jurisdictions, attuned to the aspects of policy design likely to propel them unwillingly into the spotlight. At the core of these learning processes are compelling ideas about risk: deeply persuasive notions about the extent of harm and probable futures. This study takes seriously the interpretivist contention that assessments of risk are not simple objective calculations carried out by scientific experts. Narratives about risk are socially constructed, stemming from a blend of conditions that are often economic, but also political. These narratives set regulators down particular policy pathways, in some cases tinkering around the edges of existing regulatory frameworks and in other cases engaging in wholesale reform. By charting the processes by which these ideas are translated into policy outcomes, this book has clarified a few strands of the messy and chaotic tapestry of policy making, moving us closer to understanding the dynamic politics of risk.

The Uncertain Politics of Natural Gas

In May 2023, reporters attending a world hydrogen summit in the Netherlands asked the New Brunswick incumbent Progressive Conservative premier, Blaine Higgs, to comment on the potential for "shale gas fracking" in New Brunswick. Higgs responded definitively, noting "We can not only utilize the resources we have while they are still needed, but we can offset Russian supplies, and we can shut down coal plants and convert them to a much cleaner fuel, and it would reduce emissions" (O'Connor 2023). Almost ten years after the New Brunswick debates described in this book, fracking, it seems, is again up for public discussion, at least in the eyes of the premier. As is to be expected, the economic rationale for development has endured, as have energy security claims, although this time the argument stresses the needs of Canada's European trading partners, instead of Canada's southern neighbour. Yet Higgs' comments reflect a fundamental shift in the classic economy-environment trade-off that dominated the public debates described in this book. Instead of stressing economic gains from development, the Premier's comments evoke a trifecta of economic, social, and climate benefits stemming from "very clean natural gas" (O'Connor 2023). On some level, Higgs' comments reflect the "triple win" language of sustainability and green growth common in environmental policy making (Dryzek 2021), but they also reflect the ideational transformation of hydraulic fracturing from groundwater pollutant to clean air protector. Why did the interpretive politics of natural gas change so completely in Canada since the early days of fracking politics described in this book?

One of the major changes in the landscape of the political economy of fracking since 2016 has been the rapid integration of energy and climate policy both internationally and in Canada. Historically, Canadian federal climate action has been anaemic at best, limited to largely symbolic commitments to unreachable targets based on voluntary policy measures (J. Simpson, Jaccard, and Rivers 2008; Macdonald 2020). The lack of action at the federal level continued under the pro-development mandate of the federal government under the leadership of the Conservative government and Prime Minister Harper, who was much more focused on fostering Canada's growth as an "energy superpower" rather than a climate leader (VanNijnatten and Macdonald 2020). In the absence of federal action, provincial governments moved independently on climate, with Alberta adopting output-based climate pricing in 2007, followed by British Columbia's carbon tax in 2008 and Quebec's cap-and-trade program in 2013 (Harrison 2023; 2013). During the same time period Ontario closed down its coal-fired power plants and implemented a feed-in-tariff program for renewable energy and eventually implemented a cap-and-trade program in 2017 (Stokes 2013; 2016; Millar et al. 2021).[1] Despite these gains, in the absence of federal leadership, inequities between "hydro" and "carbon" provinces grew during this time period, with British Columbia and Quebec deepening their climate commitment on the one hand and Saskatchewan and Alberta increasing oil and gas production and corresponding GHG emissions on the other (Harrison 2023; Macdonald 2020).

These norms of provincial independence, free from federal intervention, shifted dramatically however, with the election of the Liberal party to the federal government in 2015. The alignment of a federal Liberal party seeking to differentiate itself through climate leadership, together with the election of the provincial NDP in Alberta in May 2015 provided a brief window for unilateral federal action on climate (Harrison 2023; Harrison and Bang 2022). National action was reinforced by the international negotiations surrounding the Conference of the Parties to the UN Framework Convention on Climate Change (UNFCCC) in Paris in November 2015, which led to the adoption of the Paris Agreement, a protocol under the UNFCCC in which member states committed to Nationally Determined Contributions to reduce GHG emissions in the hopes of limiting global warming to a 1.5 degree centigrade rise (Dimitrov 2016). In 2016 Canada committed to a 30 per cent reduction of GHG emissions below 2005 levels by 2030, which it increased to a 40 per cent reduction in 2021 (Government of Canada 2021). In 2016 the federal government implemented, with initial agreement from all the provinces except Saskatchewan, the Pan Canadian Framework on Climate Change (PCF), a wide-ranging plan of legislative, regulatory, and policy reform designed to reduce Canada's GHG emissions and work towards meeting its commitments under the Paris Agreement (Harrison 2023; ECCC 2016).

At the heart of the PCF was a multifaceted carbon pricing scheme that built on the existing carbon pricing program in place in Alberta, British

Columbia, Quebec, and Ontario (the "carrot") while also establishing a federal equivalency benchmark that all provinces would have to meet or be subject to a federal administered carbon price (the "stick"). The federal price, or "backstop" was legislated in 2018 and is comprised of two core elements, namely a fuel charge, colloquially known as a carbon tax, and an "output-based pricing system" for industry emitters designed to lessen the impact of the carbon price on energy-intensive, trade-exposed industries (Harrison 2023). Subsequent Conservative governments in Alberta, Saskatchewan, Ontario[2] legally challenged the federal governments' jurisdiction to implement the backstop; in 2021 the Supreme Court of Canada upheld the federal government's jurisdiction to implement a minimum carbon price in the national interest (Harrison 2023).

The unprecedented climate action of the federal government and the resistance of the provincial provinces reflects broader "climatization of global politics" (Cantoni et al. 2023) that has intensified since 2016. Commitments made under the Paris Agreement have encouraged governments to adopt a climate lens through which to assess the costs and benefits of energy policy, leading to increased investments in renewable energy, electric transportation, as well as commitments to phase-out coal-fired power. At the same time, the increased attention of governments to climate action has also elicited intensive resistance among the more energy dependent states to policies likely to initiate deep decarbonization, reflected in the ongoing politics of climate denial and delay (Cantoni et al. 2023, 2; Lamb et al. 2020; Seto et al. 2016). In Canada, climate concerns have reinvigorated provincial debates around the need for and efficacy of pipelines (Janzwood 2020; 2021; Hoberg 2021), integration of provincial electricity grids (Kanduth and Dion 2022; Pineau 2021; Shaffer 2021), and investment in small modular nuclear reactors (Bratt 2021; 2020). The upshot of the increased salience of climate outcomes is that energy policy makers face an even more complex trade-off between local economic benefits (or perceived harms, in the case of stranded assets), local and global climate benefits, and local environmental harms (Neville 2021).

Within this context, natural gas, and by extension, shale gas and fracking, continues to hold an ambiguous and uncertain interpretive space. In the United States, renewed attention to the climate crisis as a policy problem has provided pro-development actors, including both state governments and industry members, with an opportunity to bolster clean energy frames to justify shale gas production, including championing natural gas as an alternative to coal-fired electricity (Neville 2021). In this clean energy frame, natural gas from domestic fracking is framed as a necessary step towards decarbonizing electricity grids by providing a firm, reliable source of electricity generation that releases lower carbon emissions in comparison to coal-fired power. This framing positions natural gas as transitional "bridge" between past dependence on coal-fired power and future reliance on renewable energy (Delborne et al. 2020). Natural gas is thus situated as a necessary evil that provides short-term economic benefits to industries and consumers while mitigating the environmental harms of oil and coal. Increased attention of governments to climate goals in North America has corresponded with an increase in the prevalence of these "bridge frames" in public debate, shifting North American discourse closer to that of pro-development European counterparts, who were more likely to refer to clean energy benefits in the earlier days of fracking debates, in part due to more stringent climate policies in the European Union (Bomberg 2017b; Chen 2020b; Delborne et al. 2020; Goldthau and Sovacool 2016; Mildenberger 2020).

On the other hand, climate activists (and social scientists) have noted that bridge frames promote a type of "fossil fuel solutionism" that reinforces climate delay, effectively slowing down the pace of policy action aimed at decarbonization (Lamb et al. 2020). Counter narratives focus on the climate costs of natural gas, namely how natural gas extraction and use prolongs societal dependence on fossil fuels, exacerbating climate change and reinforcing carbon lock-in (Brauers, Braunger, and Jewell 2021; Kemfert et al. 2022; Buschmann and Oels 2019; Brauers 2022). Again, this type of attention to climate harms was initially more prevalent in the European, rather than North American context, as anti-fracking activists in the early part of the 2010s sought to connect with electorates attuned to climate action and governments seeking to distinguish themselves as climate leaders (Bomberg 2017b). Since 2016, climate activists in Europe have sought to build on anti-fracking moratoria by extending bans more broadly to cover all fossil fuel extraction, based on a "keep it in the ground" (KIIG) rationale, evident in social movement activism in Ireland, Spain, and Germany (Carter and McKenzie 2020; McKenzie and Carter 2021). Research on the UK case has documented an increase in attention to carbon lock-in among anti-fracking coalitions, with activists drawing attention to the negative environmental consequences of fugitive methane emissions generated by fracking (Williams and Sovacool 2019). Bradshaw et al. (2022) argue that post Paris Agreement, the enthusiasm of the UK government for shale gas development was increasingly seen as counter to its climate ambitions and reinforced perceptions among the electorate that shale gas development was unnecessary in the country, a narrative that served to lay the groundwork for the 2019 moratorium.

In the Canadian context, discourse regarding climate benefits and/or climate harms has served to entrench, rather than disrupt, regulatory variation in the country. By and large, fracking moratoria and bans established in the mid 2010s have been upheld or reaffirmed, with Quebec, Newfoundland and Labrador, and Nova Scotia all reiterating restrictions on the practice in the years from 2016 to 2018 (Kestler-D'Amours 2018; Laroche 2019; L. Bird 2016). New Brunswick is the one jurisdiction that has attempted to rollback its precautionary stance, with the Higgs' government passing an order-in-council in 2019 to partially revoke the moratoria in an attempt to bolster new investment near Sussex (Poitras 2019a). However the only oil and gas firm based in the province, Corridor Resources, has chosen not to pursue development, citing "regulatory uncertainty" regarding Indigenous consultation as a stumbling block to future investment (Magee 2019). Despite Premier Higgs' attempt to reopen the political debate regarding fracking in the province in 2022, especially in the wake of the current Ukraine-Russia war, at the time of writing there has been little uptake of either the bridge fuel narrative or energy security frames in public discourse in New Brunswick. On the other hand, despite the successes of KIIG narratives and supply-side policies in Europe, debates regarding carbon lock-in continue to fall flat in the Canadian context. Although a few national environmental organizations, such as Council of Canadians, and some Atlantic organizations have advocated for a KIIG strategy, the majority of ENGOs in Canada have instead focused their attention on mitigating methane emissions, rather than attempting to halt oil and gas production in the country entirely (Janzwood and Millar 2022).

The struggle to gain political uptake around fossil fuel bans speaks to the strength of incumbent oil and gas industry in Canada but also the prevalence of triple win language that situates economic development of oil and gas as intrinsically (and paradoxically) intertwined with climate leadership (Harrison and Bang 2022). In western Canada, bridge fuel frames have served to further entrench natural gas production and use. In Alberta, both policy makers and industry acknowledge the need for energy transitions but consistently promote natural gas as the most viable solution to reducing GHG emissions. For example, Alberta's 2020 Natural Gas Vision and Strategy positions natural gas as "low-cost and low carbon" fuel providing both economic development and climate benefits (Government of Alberta 2020). In British Columbia, the framing of LNG as a path to reduced global emissions has firmly solidified from the early days of shale gas development. Its "global bridge fuel" frame justifies the extraction of "clean" natural gas in Canada because it offsets the use of coal in other jurisdictions, reducing global emissions

(Janzwood and Millar 2022; Chen 2020a). This bridge fuel rationale is a pillar of the BC government's 2016 Climate Plan, with the government stressing that "we are ensuring that we develop industries like liquefied natural gas in ways that are cleaner than competing jurisdictions, allowing us to ship it to other nations where it can reduce their reliance on higher carbon energy sources like coal and oil" (Government of British Columbia 2016, 3). Similar to the earlier discussions in Alberta regarding "ethical oil," and current debates in California regarding "clean oil" (Duffy 2023), the clean LNG narrative paradoxically acknowledges an urgent need for climate action, but uses that urgency to justify side-stepping local environmental concerns in the face of global climate benefits and energy security. In some sense, the BC bridge fuel frames illustrate a return to "triple-win" language that permeates sustainability narratives in environmental politics (Duffy 2023; Dryzek 2021). Rather than posing a trade-off between environmental benefits and environmental costs, bridge fuel narratives evoke a healthy soup of economic, social, and environmental benefits, with little-to-no environmental costs.

From a risk perspective, bridge fuel frames shift shale gas back from the edges of catastrophic risk into the relative position of linear risk, reflecting a return to the earlier perceived security of conventional gas production. In the light of competing catastrophic harms of coal-fired power, such as increasing rates of asthma among children from volatile organic compounds, or costs of climate change from GHG emissions, the environmental harms of natural gas production have been positioned as less definite, more ambiguous, and ultimately more tolerable for society by government officials and industry representatives. As such, bridge frames reinforce existing linear risk narratives in British Columbia and Alberta regarding natural gas production, and by extension, shale gas and fracking.[3] Although bridge frames acknowledge and in some cases reinforce the legitimate environmental stated goal of reducing GHG emissions, these frames also diminish the salience of local environmental harms relative to climate benefits, and in doing so, strengthen the justification for ongoing fossil fuel extraction and production.

Although climate benefits and costs seem primarily to have reinforced existing interpretive dynamics in the country, the regulation of methane emissions and other short-lived climate pollutants from oil and gas production has been an area of active regulatory development in the country post 2016. In his comparative analysis of methane policy in Canada, the United States, and Mexico, Rabe (2022) finds that following a North American summit to reduce methane emissions, the Canadian federal government began to develop federal standards for methane, including more stringent requirements for leak detection, restrictions on venting and flaring, and phasing out of high-emitting pneumatic devices in production. The federal standards were released in 2018, however, as with climate policy more generally, implementation has required the cooperation of the provincial governments and regulators. Interestingly, British Columbia has led the western provinces in developing a more comprehensive suite of regulatory tools, including a new levy designed to generate adequate revenue for remediation of orphan gas wells in the province (Rabe 2022). British Columbia's regulatory action on methane reflects a shift from its earlier "single-issue" regulation approach to a more comprehensive suite of policy instruments, akin to Colorado's regulatory approach discussed in chapter 8 (Rabe 2022). From a risk narrative perspective, the shift of the government to a more comprehensive regulatory approach suggests an increased predominance of a *complex* risk frame among bureaucrats and regulators regarding how best to reduce emissions from natural gas production. This narrative moves beyond linear risk by acknowledging the potential harm of GHG emissions but turns to experts to reduce uncertainty. Evidence suggests that BC regulators have engaged in highly technical processes of epistemic learning to reduce GHG emissions in production so as to position BC gas as "climate-friendly" on global markets (Rabe 2022).

The framing of natural gas as simultaneously clean and dirty, as the mitigator or perpetrator of climate harms, illustrates the ongoing mutability of natural gas, and nested within it, hydraulic fracturing, in energy transitions. Despite significant consensus in epistemic communities regarding the need to discontinue new investment global oil and gas production (International Energy Agency (IEA) 2021), natural gas continues to play a part in pathway scenarios, whether through provision of firm baseload power for electricity grids, or production of "low-carbon" hydrogen for the transport sector (Dion et al. 2021). In contrast to the interpretive politics of fracking, which turned on debates about immanent, certain, catastrophic environmental harms, the interpretive politics of natural gas within a climate context are rooted in debates about the speed and pace of energy transitions (Janzwood and Millar 2022; Beck and Richard 2020). Global carbon lock-in is evident in the complex web of interconnection between local production, international transportation, and global use of fossil fuels, underpinned by complex chains of finance and investment (Neville 2021; Seto et al. 2016; Buschmann and Oels 2019). Yet climate politics also draw our attention to the global fractal in which we live, namely our extensive fossil fuel use through interlocking political, economic, and social scales (Bernstein and Hoffmann 2019). For scholars and practitioners working towards decarbonization, this global fractal also signals a potential pathway to radical change, as transformation in one element of the fractal has the potential to unleash a cascade of change, shifting society to a new system at a rapid pace (Bernstein and Hoffmann 2019; Levin et al. 2012). Whether natural gas is a catalyst for rapid change or a drag on clean innovation is uncertain. Although the expansion of natural gas in electricity systems might be a short pitstop on the way to the integration of renewable energy sources in provincial grids, it could also be a lengthy visit, entrenching regions in an improved, but never fully decarbonized trajectory (Bernstein and Hoffmann 2018). The ambiguous position of natural gas as both an accelerant and/or brake in energy transitions suggests that the production, transportation, and use of gas will be an ongoing site of interpretive contestation in climate politics moving forward, both within Canada and beyond.

Within this context, close attention to the politics of risk can help identify sites of change. The case studies in this book demonstrate that risk narratives infused with dread, namely local, involuntary, and immediate harm, will inevitably generate public contention and potential for radical change. Whether this contention can scale up, entrench, and ripple across provinces and nations depends upon the deeper integration of local environmental harms with global climate costs (Carter and McKenzie 2020; Neville 2021). Alternatively, interpretive battles that pitch local environmental harms against climate benefits – whether wind farms in Ontario (Millar et al. 2021), small modular reactors in New Brunswick (O'Donnell et al. 2023), or high voltage transmission lines in Maine (Hoberg 2021) – are likely to get mired in epistemic unknowns and regulatory delay. For so many of us, uncertainty is volatile, urgent, and unnerving, generating trepidation and an unwillingness to act, both in our daily lives and our communities. But within our political systems, uncertainty is a resource, waiting to be tapped. Whether we use it to power the status quo or spark change is up to us.

List of Actor Types

Elected official (government)
Elected official (opposition)
Environmental advocate
Industry representative
Academic or scientific expert
Bureaucrat
Other resident
Municipal government official
Journalist
Indigenous nation
Health care professional
Other governments (other jurisdictions)
Oil and gas worker

Interview Schedules

British Columbia Interview Questions:

Describe your work as it relates to natural gas/energy policy in your province.

When did you and/or your agency become aware that the technologies of hydraulic fracturing and/or horizontal drilling could be used to access shale oil or gas in the province?

Where/from whom did you initially acquire information about unconventional oil and gas development? Did this change over time?

What did you identify as the key gaps or risks in the existing regulatory framework in BC (pre-2010–12)? What were the key aims of the 2010 update of the Oil and Gas Activities Act?

Has your agency consulted with outside jurisdictions to develop new regulations? If so, how and with which jurisdictions?

Which jurisdictions did you find most useful? Why?

Did your assessment of potential risks change after meeting with other jurisdictions? If so, in what areas?

How did your agency determine when it was appropriate to develop new regulation and when did the framework rely existing rules for conventional oil and gas development?

What were the key environmental issues that the proposed regulatory framework aims to address? Have these concerns changed over time (e.g., between 2008 and 2014)?

What were the key concerns raised by industry with regard to BC's royalty regime? Were there other concerns?

What were the key concerns raised by other stakeholders (e.g., landowners / municipalities / first nations) regarding the current regulatory framework?

Were there other social or economic issues that emerged that have been incorporated into the final framework?

Does your province have any environmental or health issues related to hydraulic fracturing operations which need future work/research?

Have any other provinces consulted with your agency regarding their policy process in developing regulation?

Alberta Interview Questions:

Describe your work as it relates to natural gas / energy policy in your province.

When did you and/or your agency become aware that the technologies of hydraulic fracturing and/or horizontal drilling could be used to access shale oil or gas in the province?

Where/from whom did you initially acquire information about unconventional oil and gas development? Did this change over time?

What did you identify as the key gaps or risks in the existing regulatory framework in Alberta (pre-2011) (e.g., well casing, cement bonding, water licensing, wastewater transportation and/or storage, baseline testing?)

Has your agency consulted with outside jurisdictions to develop new regulations? If so, how and with which jurisdictions?

Which jurisdictions did you find most useful? Why?

Did your assessment of potential risks change after meeting with other jurisdictions? If so, in what areas?

How did your agency determine when it was appropriate to develop new regulation and when did the AER rely on existing rules for conventional oil and gas development?

What were the key environmental issues that the proposed regulatory framework aims to address? Have these concerns changed over time (e.g., between 2010 and 2014)?

What were the key concerns raised by industry with regard to Alberta's royalty regime? Were there other concerns?

What were the key concerns raised by other stakeholders (e.g., landowners / municipalities) regarding the regulatory framework?

Were there other social or economic issues that emerged that have been incorporated into the final framework?

Does your province have any environmental or health issues related to hydraulic fracturing operations which need future work / research?

Have any other provinces consulted with the Alberta government regarding their process in developing regulation?

What communication tools do you use, if any, to educate the public about hydraulic fracturing? Who is responsible for communication about unconventional gas development?

New Brunswick Interview Questions:

Describe your work as it relates to natural gas/energy policy in your province.

When did you and/or your agency become aware that the technologies of hydraulic fracturing and/or horizontal drilling could be used to access shale oil or gas in the province?

Where/from whom did you initially acquire information about unconventional oil and gas development? Did this change over time?

What did you identify as the key gaps or risks in the existing regulatory framework in New Brunswick (pre-2011) (e.g., well casing; cement bonding; water licensing; wastewater transportation and/or storage; baseline testing?)

Did your agency consult with outside jurisdictions to develop new regulations? If so, how and with which jurisdictions? How were Pennsylvania, Arkansas, British Columbia, and Alberta selected?

Which jurisdictions did you find most useful? Why?

Did your assessment of potential risks change after meeting with other jurisdictions? If so, in what areas?

How did your agency determine when it was appropriate to develop new regulation and when did the framework rely on rules similar to existing regulations in other provinces (e.g., references to Alberta's ERCB Directives)?

What were the key environmental issues that the regulatory framework aimed to address? Did these concerns change over time (e.g., between 2010 and 2014)?

What were the key concerns raised by industry with regard to the feasibility of the casing regulations and/or the royalty regime? Were there other concerns?

What were the key concerns raised by other stakeholders (e.g., landowners / municipalities) regarding the regulatory framework?

Were there other social or economic issues that emerged that were incorporated into the final framework?

Does your province have any environmental or health issues related to hydraulic fracturing operations which need future work/research?

Have any other provinces consulted with the New Brunswick government regarding their process in developing regulation?

Nova Scotia Interview Questions:

Describe your work as it relates to unconventional gas / hydraulic fracturing / horizontal drilling in the province.

When did you or your agency initially become aware that technologies of hydraulic fracturing and/or horizontal drilling could potentially be used to access shale oil or gas in the province?

Where or from whom did you or your agency initially acquire information about oil and gas development? Did this change over time?

Did your agency participate or engage in developing policy regarding unconventional oil and gas development in your province? How?

Did your agency consult with outside jurisdictions to develop its initial policy position? If so, how and with which jurisdictions?

Which jurisdictions did you find the most useful? Why?

Have there been any events in your province or elsewhere in Canada and/or the United States which have drawn media attention to this policy area? If so, how did your department respond?

What communication tools do you use, if any, to educate the public about hydraulic fracturing and the issues being raised? Who is responsible for communication about unconventional gas development?

What are the key messages regarding hydraulic fracturing in your province that your organization aims to communicate?

What are the types of concerns being raised by landowners, communities, and interest groups in your province?

What were the initial challenges you have identified regarding unconventional oil and gas development, if any? Has your understanding change over time?

Have you identified other economic, social, or environmental issues related to hydraulic fracturing over the last four years? Have different issues become higher priority over time?

Does your province have any environmental or health issues related to hydraulic fracturing operations which need future work/research? If yes, what plans you have in this regard?

Notes

1. Fracking and the Politics of Risk

1 This study uses the term "hydraulic fracturing" to refer to the industrial process of multi-stage hydraulic fracturing combined with horizontal drilling, a process which is sometimes also referred as High Volume Hydraulic Fracturing (HVHF) (Arnold and Neupane 2017). Technical and industry experts often distinguish between the act of hydraulic fracturing – that is, the specific process of forcing volumes of water down wells to fracture rock formations and release natural gas or tight oil – and other activities of drilling, transportation, and site management that surround the process (CAPP 2017). The study uses hydraulic fracturing as shorthand for the entire process of development.

2 Policy communities, or subsystems include a range of policy elites, including bureaucrats, government officials, environmental advocates, industry representatives, journalists, academics, whose shared beliefs – including both policy preferences and deep core beliefs – can potentially inform and/or shape regulatory design (Sabatier 1988; Skogstad 2008; Howlett, Ramesh, and Perl 2009).

3 I omit Ontario and Quebec from the core analysis, the former because of the very nascent policy development and the latter because of its *sui generis* position within the Canadian federation, as well as Saskatchewan and Newfoundland mainly because hydraulic fracturing in these two provinces is focused on tight or shale oil, rather than shale gas. See Millar (2020; 2021) for initial case comparisons developed further in this book.

2. Fracking Uncertainty

1 The practice described below is often also termed High Volume Hydraulic Fracturing in the scholarly literature (Arnold and Neupane 2017).

2 BTU refers to "British Thermal Unit," a measurement for heat content of energy sources used in North American markets. One thousand cubic feet (Mcf) of natural gas equals approximately 1.037 million Btu (MMBtu) (US EIA 2018b).

3 Henry Hub is a natural gas pipeline that sets the price for trading gas futures contracts on the NYMEX. As such spot prices at Henry Hub function as a proxy for US gas prices.

4 A resource "play" is a term used to describe the wide distribution of natural gas resources across an extensive geographic area (NEB 2009, 5).

5 The Barnett shale in Texas, one of the earliest plays to be developed in the United States, is estimated at 44 marketable Tcf (NEB and BCMEM 2011). The Canadian domestic market demand for natural gas is approximately 1 Tcf annually.

6 It is important to note that the specific process of hydraulic fracturing (e.g., using volumes of water to fracture rock) has been used on conventional wells in Alberta since the 1950s (AEP 2018); however the combination of horizontal drilling and high volume has only become widespread in the province since 2013 (Natural Resources Canada 2015).

7 Industry calculates marketable resources at 5 to 20 per cent of gas in place (GIP) (Rokosh et al. 2009).

8 Studies in the United States have identified a range of 13,700 to 23,800 cubic metres per well (Kondash and Vengosh 2015).

9 For example, "fracfocus.org" and its partner site "fracfocus.ca" are maintained through a collaboration of industry and environmental groups: the Ground Water Protection Council and Interstate Oil and Gas Compact Commission in the United States and the BC Oil and Gas Commission in Canada (Konschnik and Dayalu 2016).

10 Rabe's findings also suggest that researchers need to consider the role of historical regulatory systems for oil and gas in relation to hydraulic fracturing regulations.

11 To date, comparative social science research conducted on Canadian provincial regulations has been relatively limited. For exceptions, see Stephenson and Shaw (2013) and Carroll et al. (2012) for early case studies of British Columbia; Montpetit, Lachapelle and Harvey (2016) and Montpetit and Lachapelle (2017) for a comparison of British Columbia and Quebec; Carter and Eaton (2016) for a brief overview of Canadian regulation and an in-depth case study of regulation in Saskatchewan; Gagnon et al. (2015) for a discussion of water quality regulations in Alberta, British Columbia, and New Brunswick; and Neville and Weinthal (2016) for a discussion of public consultations regarding hydraulic fracturing in the Yukon.

12 The study reviews these four provinces because during the analytic period (2006–16) they were the provinces with the most developed regulatory frameworks for shale gas and tight oil in Canada (Energy and Mines Minister's Conference 2013b; Council of Canadian Academies 2014). Manitoba and PEI have yet to develop specific regulations for hydraulic fracturing, relying on existing oil and gas drilling regulations developed for conventional production (Natural Resources Canada 2015). In 2013 Ontario reported working on an internal review of its regulatory framework but did not released a public report (Energy and Mines Minister's Conference 2013b; Council of Canadian Academies 2014; Natural Resources Canada 2017).

13 Carter and Eaton also similarly categorize this variation in their study of Saskatchewan (2016), drawing on Rabe and Borick (2013).

14 The Gallant government reaffirmed its position by indefinitely extending the moratorium in May, 2016 (McHardie 2016). In June 2019 the subsequent government passed an order in council to exempt certain parcels of land from the moratorium (Canadian Association of Physicians for the Environment, Macfarlane, and Perrotta 2020; Global News 2019).

3. Analysing Uncertainty

1 This study is aligned with Clark's (2013) discussion of "risk frames."

2 Economic classifications of risk tend to focus the extent of damage and probability of occurrence (Zinn 2008; Falkner and Jaspers 2012). Similar to economic concepts of insurance, technical risk assessments determine the expected value of a particular risk by calculating the degree of potential harm estimated or observed using scientific methods and multiplying it by probability of occurrence, often determined through statistical modelling based on historical data (Zinn 2008). Despite the predominance of this conception of risk in technical risk assessment (Renn 2008), a key finding from cognitive psychology is the importance of risk perception, or the social construction of risk, in determining individual risk assessments.

3 Research suggests that this relationship between beliefs and support / opposition may not be unidirectional; in a national US survey Evensen and Stedman (2017) find support for the hypothesis that general attitudes of support or opposition for hydraulic fracturing drive beliefs about environmental risks and harms.

4 Readers will note that the focus of this study on the causal influence of risk narratives in the policy process by and large side-steps the issue of what elements of the policy-making context, and particularly the political economy, makes the emergence of certain risk narratives more or less likely. I return to this point in chapter 8.

5 By and large, scholarship on policy learning has focused on the activities of policy elites, examining the full range of actors within a policy subsystem, including political decision makers and government officials as well as interest group representatives, academics, and journalists (May 1992; Bennett and Howlett 1992; Pierson 1993; Sabatier and Weible 2007; Hoberg 1996; Jenkins-Smith 1988).

6 In general the study follows Sabatier and Weible's (2007) suggestion that policy subsystems be examined for at least ten years to capture policy change. The research focuses primarily on the years from 2010 to 2014, when the majority of policy debate occurred; however in British Columbia the regulatory and media analysis extend earlier to 2003 in order to capture early policy developments that the government engaged in as a first mover in the field. Scholars examining hydraulic fracturing have suggested that the policy area can be considered emerging, and as such is a nascent, rather than fully mature subsystem, making the compressed timeline appropriate (Ingold, Fischer, and Cairney 2016).

7 In addition, the political dynamics of regulatory development have been deftly examined through Angela Carter's work on petro-provinces; my study serves as a complement to her groundbreaking work examining environmental regulatory development in Saskatchewan, Alberta, and Newfoundland and Labrador (Carter 2020; Carter and Fusco 2017; Carter and Eaton 2016; Carter, Fraser, and Zalik 2017).

8 The BAPE is a provincial government agency designed to facilitate public engagement on the ecological, social, and economic impacts of projects on citizens by providing policy advice to Minister of Sustainable Development, Environment, and the Fight against Climate Change. The BAPE provides an institutionalized forum for public information sessions, public hearings, as well as facilitating mediation between project proponents and citizens (BAPE 2018).

9 See appendix 1 for a full list of actor types.

4. Limiting Uncertainty in British Columbia

1 The government was led by the BC Liberal Party from 2001 to 2017.

2 Google trends data indicates a peak in interest in searches of hydraulic fracturing in November 2011 https://trends.google.com/trends /explore?date=all&q=hydraulic%20fracturing.

3 By September 2009 the domestic spot price for natural gas had fallen to $3 MBtu from highs of over $12 MBtu in 2008 (Morrissy 2009; Foss 2011); since 2009 prices have vacillated from below $2.00 MBtu to slightly above $6.00, reflecting the overall over-supply of shale gas stemming from US production (US EIA 2018a).

4 The first stage of the assessment was conducted by Fraser Basin Council and entailed a public consultation on scope conditions for the assessment (Fraser Basin Council 2012).

5 Subsequent analysis of these projections (Hughes 2015; M. Lee 2015) has highlighted the highly variable and uncertain nature of resource estimates and called into question the reliability of the methods used to estimate the number of permanent and temporary jobs. Hughes (2015) estimates that the high export case (5 LNG export terminals) would require 1,435 wells to be drilled per year from 2022 to 2040 to support production for export. Existing estimates of marketable reserves would need to increase four-fold to meet this demand. In the absence of an LNG export industry, Hughes projects drilling rates of 292 wells/year from 2022 to 2040 to maintain current levels of production (Hughes 2015, 27).

6 Analysis of information gathered under the British Columbia Lobbyist Registration Act conducted by Graham et al. (Graham, Daub, and Carroll 2017) finds that between 2010 and 2016 ten of the most active oil and gas companies accounted for 75 per cent of the lobbying conducted by the sector, reflected in over 19,000 contacts with government officials in the six year time frame.

7 The IOGCC is a network of thirty US states and nine international affiliates, including Alberta, British Columbia, Saskatchewan, New Brunswick, Nova Scotia, Newfoundland and Labrador, and Yukon (IOGCC, n.d.; 2018)

8 It is relevant to note that LNG can be generated by conventional gas production in addition to fracking, and so these terms are not synonymous. However a majority of proposed BC LNG export projects are supplied by hydraulic fracturing in northeastern British Columbia (Heerema and Kniewasser 2017).

9 That proximity is an important factor in activating dread risk is evident in the significant mobilization in the Lower Mainland of British Columbia against the Trans Mountain Pipeline Expansion project, the terminus of which was located in Burnaby, the urban centre adjacent to the City of Vancouver (Hoberg 2021, 124).

5. Monitoring Uncertainty in Alberta

1 There are five articles in the dataset referring to Ernst's case.

6. Managing Uncertainty in New Brunswick

1 It is less clear whether the intermediate stance of these actors is based on a lack of scientific consensus regarding the causal risks of hydraulic fracturing or is the result of journalistic norms or a combination thereof.

2 The Natural Gas Group also conducted a consultation with representatives from government, industry, academia, First Nations, and environmental groups in June 2011 (New Brunswick Natural Gas Group 2012b). See below for a discussion of the impact of the 2011 forum on the anti-fracking coalition.

3 The discussion paper was released with a companion document of references that is 36 pages long, including references to legislation and regulation in Alberta, British Columbia, Colorado, Delaware, Maryland, Michigan, Montana, New Jersey, New Mexico, New York, Nova Scotia, Pennsylvania, Philadelphia, Quebec, Saskatchewan, Texas, West Virginia, and Wyoming.

4 *Gasland* is a feature length film focusing on the contamination of wells in local communities in Pennsylvania and the growth of fracking in the US oil and gas industry (Fox 2010). Research finds that the film generated considerable attention in both the US and Canadian contexts by providing an opportunity for activists to highlight environmental and health risks (Vasi et al. 2015; Eaton and Kinchy 2016).

5 Interestingly, this strategy did not prevent Gallant from asserting that science would be able to resolve the debate in the long term, following the maturation of the field.

7. Contesting Uncertainty in Nova Scotia

1 The study codes direct quotes to specific actor types; unattributed quotes are assigned to the journalist category.

2 In comparison, the study coded half the number of references to only eight other jurisdictions in the New Brunswick dataset.

3 The broader coverage could reflect the increasing salience of the policy issue in other jurisdictions or may simply be a function of the preferences of the Herald editorial board.

4 Parker replaced Bill Estabrooks who stepped down from the position of Energy Minister due to health concerns.

5 The ERCB in Alberta also conducted a jurisdictional scan which it released in January 2011; the report reviewed regulatory regimes in British Columbia, Saskatchewan, Louisiana, Michigan, New York, Oklahoma, Pennsylvania, and Texas (ERCB 2011).

6 The potential for public meetings to increase, rather than dampen, salience was not lost on bureaucratic advisers within government who strongly recommended against holding public meetings (confidential interviewee 2014i).

7 Gallant did however draw on public concern as justification for the governments' decision to uphold the moratorium in 2016.

8. Regulating Uncertainty

1 Responsibility for regulation of hydraulic fracturing has also been devolved to national governments, which has led to variation across the United Kingdom. Northern Ireland declared a moratorium in 2015, followed by Scotland in 2017, and Wales in 2018 (Evensen 2018; Stephan 2020; Government of Wales 2018).

2 In a separate vein, future research could consider the relative influence of geography on the emergence of risk narratives. As the British Columbia case demonstrated, one of challenges cited by environmental activists in the province was the perceived distance from the (rural) site of energy extraction and urban habitation. Interviewees noted that perceived distance made it difficult to mount province-wide campaigns on local impacts such as health risks, or groundwater contamination. Conversely, a "not-in-my-backyard" (NIMBY) (Wolsink 2007) approach could perhaps explain some of the public resistance to development in New Brunswick, where the shale play is much closer to population centres than in British Columbia. However, an emerging body of literature in environmental politics argues that there is a complex and multidirectional relationship between actual distance, perceived distance, risk perception, and support for energy development (Gravelle and Lachapelle 2015; Clarke et al. 2016; Alcorn, Rupp, and Graham 2017). For example, an a 2016 public opinion survey on hydraulic fracturing conducted with residents of British Columbia and New Brunswick found that while living in an urban centre helped predict perceptions of economic benefits among New Brunswick respondents, location did not have a significant effect on perceptions of environmental risks in either province (O'Connor and Fredericks 2018).

   Moreover, there is some reason to suspect that there is a feedback effect between place, perceived identities, and the social construction of risk (Eaton and Kinchy 2016). For example, findings from cases of public opposition to wind turbine siting suggest that "disruption of place-attachment" can be a significant predictor of public opposition, as threats to physical and symbolic attributes of certain locations can activate residents' sense of rural identity in opposition to urban policy elites and energy proponents (Devine-Wright and Howes 2010, 271). A future avenue of research would be to examine the interactive relationship between risk narratives, identities, and place. For example, do changes in actors' rural / suburban / urban identities affect perceptions of risk? Or when actors have an ideological closeness to given space – as in the case of the British Columbia coast for example – does that change their dominant risk narrative?

3 The right of free, prior, and informed consent under the UN Declaration on the Rights of Indigenous Peoples also presents an avenue of political authority that has been tentatively acknowledged by the Canadian federal government, albeit within a legal perspective that asserts the dominant sovereignty of the Canadian state (Papillon and Rodon 2017; Lightfoot 2020).

4 For example, in a study of public engagement in housing policy decision making, Doberstein and Millar (2014) identified consensus decision-making rules, a deliberative ethic, and open access to procedures and decisions as features that bolstered procedural legitimacy in networked governance structures.

Epilogue

1 Ontario's cap-and-trade program was short lived; the Ontario Progressive Conservatives, led by Doug Ford, cut the program almost immediately after attaining office in the summer of 2018 (Raymond 2019; Lachapelle and Kiss 2019).

2 New Brunswick initially joined the provincial challengers, however following the second election of the federal Liberal party in 2019, the New Brunswick government announced its intention to implement a provincial carbon price (Mildenberger 2019; Poitras 2019b).

3 As noted elsewhere, this framing also serves to sidestep the GHG implications of British Columbia's healthy export industry of metallurgical coal (Janzwood and Millar 2022; Harrison 2020).

References

Adams, Christopher. 2011. "Summary of Shale Gas Activity in Northeast British Columbia 2011." Victoria: BC Ministry of Energy and Mines.

Adams, Michelle, David Wheeler, and Genna Woolston. 2011. "A Participatory Approach to Sustainable Energy Strategy Development in a Carbon-Intensive Jurisdiction: The Case of Nova Scotia." *Energy Policy* 39 (5): 2550–9. https://doi.org/10.1016/j.enpol.2011.02.022.

Alberstat, Joann. 2013. "Poll: 53% in N.S. Oppose Fracking." *Halifax Chronicle Herald*, 10 April. http://thechronicleherald.ca/business/1122653-poll-53-in-ns-oppose-fracking.

Alberta Energy Regulator (AER). 2013a. "Directive 083: Hydraulic Fracturing – Subsurface Integrity." https://www.aer.ca/documents/directives/Directive083.pdf.

– 2013b. "Alberta Energy Regulator Directors Get Down to Business with First Meeting." Alberta Energy Regulator. https://www.globenewswire.com/en/news-release/2013/06/18/1347347/0/en/Alberta-Energy-Regulator-Directors-Get-Down-to-Business-With-First-Meeting.html.

– 2014. "Frequently Asked Questions: Play-Based Regulation (PBR) Pilot Project." Alberta Energy Regulator. https://web.archive.org/web/20160326045835/https://www.aer.ca/documents/about-us/PBR_FAQ_20141204.pdf.

– 2016. "Evaluation of the Alberta Energy Regulator's Play-Based Regulation Pilot." Calgary: Alberta Energy Regulator. https://www.aer.ca/documents/about-us/PBR_EvaluationReport_June2016.pdf.

– 2018. "Organizational Structure." http://www.aer.ca/providing-information/about-the-aer/who-we-are/aer-organizational-structure#stake.

Alberta Environment and Parks (AEP). 2018. "Hydraulic Fracturing | AEP – Environment and Parks." Hydraulic Fracturing. http://aep.alberta.ca/water/water-conversation/hydraulic-fracturing.aspx.

Alcorn, Jessica, John Rupp, and John D. Graham. 2017. "Attitudes toward 'Fracking': Perceived and Actual Geographic Proximity." *Review of Policy Research* 34 (4): 504–36. https://doi.org/10.1111/ropr.12234.

ALLconsulting. 2012. "The Modern Practices of Hydraulic Fracturing: A Focus on Canadian Resources Revised November." Calgary: Petroleum Technology Alliance Canada.

Allison, Edith, and Ben Mandler. 2018. "Petroleum and the Environment." Alexandria, VA: The American Geosciences Institute. https://www.americangeosciences.org/sites/default/files/AGI_PetroleumEnvironment_web.pdf.

Alward, David. 2014. Personal interview conducted by phone 1 December.

Andersson-Hudson, Jessica, William Knight, Mathew Humphrey, and Sarah O'Hara. 2016. "Exploring Support for Shale Gas Extraction in the United Kingdom." *Energy Policy* 98 (November): 582–9. https://doi.org/10.1016/j.enpol.2016.09.042.

Andrews, Richard, and Austin Holland. 2015. "Summary Statement on Oklahoma Seismicity." Oklahoma Geological Survey. https://web.archive.org/web/20150427160628/http://earthquakes.ok.gov/wp-content/uploads/2015/04/OGS_Summary_Statement_2015_04_20.pdf.

Angen, Eli, and Jason Switzer. 2012. "Shale Gas Thought Leader Forum: Forum Proceedings." Pembina Institute. http://www.pembina.org/reports/shale-gas-thought-leader-forum-proceedings.pdf.

Angevine, Gerry, and Miguel Cervantes. 2010. "Global Petroleum Survey 2010." The Fraser Institute. https://www.fraserinstitute.org/studies/global-petroleum-survey-2010.

Arnold, Gwen, and Robert Holahan. 2014. "The Federalism of Fracking: How the Locus of Policy-Making Authority Affects Civic Engagement." *Publius: The Journal of Federalism* 44 (2): 344–68. https://doi.org/10.1093/publius/pjt064.

Arnold, Gwen, Le Anh Nguyen Long, and Madeline Gottlieb. 2017. "Social Networks and Policy Entrepreneurship: How Relationships Shape Municipal Decision Making about High-Volume Hydraulic Fracturing." *Policy Studies Journal* 45 (3): 414–41. https://doi.org/10.1111/psj.12175.

Arnold, Gwen, and Kaubin Wosti Neupane. 2017. "Determinants of Pro-Fracking Measure Adoption by New York Southern Tier Municipalities." *Review of Policy Research* 34 (2): 208–32. https://doi.org/10.1111/ropr.12212.

Atherton, Frank, Kevin Christmas, Shawn Dalton, Maurice Dusseault, Graham Gagnon, Brad Hayes, Constance MacIntosh, Ian Mauro, Ray Ritcey, and David Wheeler. 2014. "Report of the Nova Scotia Independent Review Panel on Hydraulic Fracturing." Sydney, Nova Scotia: Cape Breton University. http://energy.novascotia.ca/sites/default/files/Report%20of%20the%20Nova%20Scotia%20Independent%20Panel%20on%20Hydraulic%20Fracturing.pdf.

Atkinson, Michael M., and William D. Coleman. 1992. "Policy Networks, Policy Communities and the Problems of Governance." *Governance* 5 (2): 154–80. https://doi.org/10.1111/j.1468-0491.1992.tb00034.x.

Ayers, Tom. 2014a. "Sparks Fly at Public Session on Fracking." *Halifax Chronicle Herald*, 16 July. http://thechronicleherald.ca/novascotia/1223212 -sparks-fly-at-public-session-on-fracking.

– 2014b. "CBU President Denies Conflict in Fracking Review." *Halifax Chronicle Herald*, 1 August. http://thechronicleherald.ca/novascotia/1226957-cbu -president-denies-conflict-in-fracking-review.

Baka, Jennifer, Kate J. Neville, Erika Weinthal, and Karen Bakker. 2018. "Agenda-Setting at the Energy-Water Nexus: Constructing and Maintaining a Policy Monopoly in U.S. Hydraulic Fracturing Regulation." *Review of Policy Research* 35 (3): 439–65. https://doi.org/10.1111/ropr.12287.

Bamberger, Michelle, and Robert E. Oswald. 2016. "Impacts of Gas Drilling on Human and Animal Health." *NEW SOLUTIONS: A Journal of Environmental and Occupational Health Policy*, August. https://doi.org/10.2190/NS.22.1.e. Medline:22446060

BAPE, Bureau d'audiences publiques sur l'environnement. 2018. "BAPE-Organization." https://web.archive.org/web/20160305003347/http:// www.bape.gouv.qc.ca/sections/bape/organisme/eng_organization_ind .htm.

Baumgartner, Frank R., Suzanna L. De Boef, and Amber E. Boydstun. 2008. *The Decline of the Death Penalty and the Discovery of Innocence*. 1st ed. Cambridge; New York: Cambridge University Press.

Baumgartner, Frank R., and Bryan D. Jones. 1993. *Agendas and Instability in American Politics*. Chicago: University of Chicago.

Baumgartner, Frank R., Bryan D. Jones, and Peter B. Mortensen. 2014. "Punctuated Equilibrium Theory: Explaining Stability and Change in Public Policymaking." In *Theories of the Policy Process*, ed. Paul A. Sabatier and Christopher M. Weible. 3rd ed. Boulder, CO: Westview Press.

Baumgartner, Frank R., and Christine Mahoney. 2008. "Forum Section: The Two Faces of Framing Individual-Level Framing and Collective Issue Definition in the European Union." *European Union Politics* 9 (3): 435–49. https://doi.org/10.1177/1465116508093492.

Baxter, Joan. 2011. "Frack and Forth on Shale Gas." *Halifax Chronicle Herald*, 16 October. http://thechronicleherald.ca/thenovascotian/25395-frack-and -forth-shale-gas.

BC Oil and Gas Commission. 2011. "Industry Bulletin 2011–33: Submission Information for Disclosure of Hydraulic Fracturing Fluids." BC Oil and Gas Commission. https://www.bcogc.ca/node/6067/download.

– 2012a. "Investigation of Observed Seismicity in the Horn River Basin." Victoria, BC: BC Oil and Gas Commission.

– 2012b. "About NEWT (NorthEast Water Tool)." Victoria, BC: BC Oil and Gas Commission. https://www.bcogc.ca/sites/default/files/documentation/web-pages/about-newt.pdf.
– 2012c. "Drilling and Production Regulation Amendments (August 2012)." Victoria, BC: BC Oil and Gas Commission. https://www.bcogc.ca/node/8051/download.
– 2013. "Area-Based Analysis." Victoria, BC: BC Oil and Gas Commission. https://www.bcogc.ca/node/12265/download.
– 2014a. "About Us | BC Oil and Gas Commission." Victoria, BC: BC Oil and Gas Commission. http://www.bcogc.ca/about-us.
– 2014b. "Oil and Gas Land Use in Northeast British Columbia." Victoria, BC: BC Oil and Gas Commission. https://www.bcogc.ca/node/12908/download.
– 2014c. "Investigation of Observed Seismicity in the Montney Trend." Victoria, BC: BC Oil and Gas Commission. https://www.bcogc.ca/node/12291/download.
– 2015. "British Columbia's Oil and Gas Reserves and Production Report." Victoria, BC: BC Oil and Gas Commission. https://www.bcogc.ca/node/13607/download.
BCBC, BC Business Council. 2013. "Building New Energy Advantages for BC: Understanding and Benefitting from the Transformation of BC's Energy Marketplaces." Vancouver: BC Business Council. http://www.bcbc.com/content/997/2013%2010%2015%20Energy%20White%20Paper%20FINAL.pdf.
Beach, Derek. 2016. "It's All about Mechanisms – What Process-Tracing Case Studies Should Be Tracing." *New Political Economy* 21 (5): 463–72. https://doi.org/10.1080/13563467.2015.1134466.
Beach, Derek, and Rasmus Brun Pedersen. 2013. *Process-Tracing Methods: Foundations and Guidelines.* University of Michigan Press. https://muse.jhu.edu/book/21935.
Beck, Marisa, and Aimee Richard. 2020. "What Is 'Transition'? The Two Realities of Energy and Environmental Leaders in Canada." Ottawa, ON: Positive Energy, University of Ottawa. https://www.uottawa.ca/research-innovation/sites/g/files/bhrskd326/files/2022-08/what_is_transition_final_web.pdf.
Béland, Daniel. 2005. "Ideas and Social Policy: An Institutionalist Perspective." *Social Policy & Administration* 39 (1): 1–18. https://doi.org/10.1111/j.1467-9515.2005.00421.x.
– 2010. "Reconsidering Policy Feedback How Policies Affect Politics." *Administration & Society* 42 (5): 568–90. https://doi.org/10.1177/0095399710377444.
Béland, Daniel, Martin B. Carstensen, and Leonard Seabrooke. 2016. "Ideas, Political Power and Public Policy." *Journal of European Public Policy* 23 (3): 315–17. https://doi.org/10.1080/13501763.2015.1122163.

Béland, Daniel, and Robert Henry Cox. 2010. *Ideas and Politics in Social Science Research*. Oxford: Oxford University Press.

Bennett, Colin J., and Michael Howlett. 1992. "The Lessons of Learning: Reconciling Theories of Policy Learning and Policy Change." *Policy Sciences* 25 (3): 275–94. https://doi.org/10.1007/BF00138786.

Bernstein, Steven, and Matthew Hoffmann. 2018. "The Politics of Decarbonization and the Catalytic Impact of Subnational Climate Experiments." *Policy Sciences* 51 (2): 189–211. https://doi.org/10.1007/s11077-018-9314-8. Medline:31007288

– 2019. "Climate Politics, Metaphors and the Fractal Carbon Trap." *Nature Climate Change* 9 (12): 919–25. https://doi.org/10.1038/s41558-019-0618-2.

Berry, Shawn. 2012. "N.B. Urged to Monitor Shale Gas Health Impacts; Report from Chief Medical Officer of Health Calls for Changes before Resource Development Begins." *Moncton Times and Transcript*, 16 October. Sec. Main.

– 2013. "LaPierre Fallout Ripples to Shale Gas File." *Moncton Times and Transcript*, 20 September. Sec. A.

– 2014a. "Leaders React to N.S. Ban on Fracking." *Moncton Times and Transcript*, 4 September. Sec. Main.

– 2014b. "Liberals Outline Economic Blueprint in Campaign Platform." *Moncton Times and Transcript*, 8 September. Sec. International.

Beswick, Aaron. 2012. "LNG Plan Gives Hope to Goldboro Families." *Halifax Chronicle Herald*, 25 October.

– 2013. "Fracking Water Won't Be Sent through Sewer System in Colchester County." *Halifax Chronicle Herald*, 17 May. http://thechronicleherald.ca/novascotia/1130072-fracking-water-wont-be-sent-through-sewer-system-in-colchester-county.

Bherer, Laurence, Pascale Dufour, and Christine Rothmayr Allison. 2013. "Public Mobilisation, Delegation,and Policy Change: The Case of Shale Gas in Quebec." Presented at the annual conference of the Canadian Political Science Association. Victoria, BC.

Bird, Geoff. 2011. "Province Releases Natural Gas Rules; Government Hoping for More Accountability in Shale Gas Industry with New Regulations." *Moncton Times and Transcript*, 24 June. Sec. International.

Bird, Lindsay. 2016. "Fracking Not 'a Game Changer' for N.L., Says Independent Report." *CBC News*, 31 May. http://www.cbc.ca/news/canada/newfoundland-labrador/western-nl-hydraulic-fracturing-report-released-1.3607408.

Birkland, Thomas A. 1998. "Focusing Events, Mobilization, and Agenda Setting." *Journal of Public Policy* 18 (1): 53–74. https://doi.org/10.1017/S0143814X98000038

Bissett, Kevin. 2010. "N.B. Premier Calls off $3.2-Billion NB Power Sale to Hydro-Quebec." *Montreal*, 24 March. https://montreal.ctvnews.ca/n-b-premier-calls-off-3-2-billion-nb-power-sale-to-hydro-quebec-1.495114.

Black, Paul. 2014. Personal interview conducted 25 November in Halifax, Nova Scotia.

Blyth, Mark. 2001. "The Transformation of the Swedish Model: Economic Ideas, Distributional Conflict, and Institutional Change." *World Politics* 54 (1): 1–26. https://doi.org/10.1353/wp.2001.0020.

– 2002. *Great Transformations: Economic Ideas and Institutional Change in the Twentieth Century*. Cambridge: Cambridge University Press.

– 2006. "Great Punctuations: Prediction, Randomness, and the Evolution of Comparative Political Science." *American Political Science Review* 100 (4): 493–8. https://doi.org/10.1017/S0003055406062344.

– 2007. "Powering, Puzzling, or Persuading? The Mechanisms of Building Institutional Orders." *International Studies Quarterly* 51 (4): 761–77. https://doi.org/10.1111/j.1468-2478.2007.00475.x.

– 2009. "Coping with the Black Swan: The Unsettling World of Nassim Taleb." *Critical Review* 21 (4): 447–65. https://doi.org/10.1080/08913810903441385.

— 2013. "Paradigms and Paradox: The Politics of Economic Ideas in Two Moments of Crisis." *Governance* 26 (2): 197–215. https://doi.org/10.1111/gove.12010.

– 2016. "The New Ideas Scholarship in the Mirror of Historical Institutionalism: A Case of Old Whines in New Bottles?" *Journal of European Public Policy* 23 (3): 464–71. https://doi.org/10.1080/13501763.2015.1118292.

Boersma, Tim, and Corey Johnson. 2012. "The Shale Gas Revolution: U.S. and EU Policy and Research Agendas." *Review of Policy Research* 29 (4): 570–6. https://doi.org/10.1111/j.1541-1338.2012.00575.x.

Bomberg, Elizabeth. 2013. "The Comparative Politics of Fracking: Networks and Framing in the US and Europe." SSRN Scholarly Paper ID 2301196. Rochester, NY: Social Science Research Network. http://papers.ssrn.com/abstract=2301196.

– 2017a. "Shale We Drill? Discourse Dynamics in UK Fracking Debates." *Journal of Environmental Policy & Planning* 19 (1): 72–88. https://doi.org/10.1080/1523908X.2015.1053111.

– 2017b. "Fracking and Framing in Transatlantic Perspective: A Comparison of Shale Politics in the US and European Union." *Journal of Transatlantic Studies* 15 (2): 101–20. https://doi.org/10.1080/14794012.2016.1268789.

Boothe, Katherine, and Kathryn Harrison. 2009. "The Influence of Institutions on Issue Definition: Children's Environmental Health Policy in the United States and Canada." *Journal of Comparative Policy Analysis: Research and Practice* 11 (3): 287–307. https://doi.org/10.1080/13876980903220736.

Boswell, Christina. 2009. *The Political Uses of Expert Knowledge: Immigration Policy and Social Research*. 1st ed. Cambridge: Cambridge University Press.

Bott, Robert. 2004. "Evolution of Canada's Oil and Gas Industry." Calgary: Canadian Centre for Energy Information. http://www.energybc.ca/cache/oil/www.centreforenergy.com/shopping/uploads/122.pdf.

Boudet, Hilary. 2019. "Public Perceptions of and Responses to New Energy Technologies." *Nature Energy* 4 (6): 446–55. https://doi.org/10.1038/s41560 -019-0399-x.

Boudet, Hilary, Dylan Bugden, Chad Zanocco, and Edward Maibach. 2016. "The Effect of Industry Activities on Public Support for 'Fracking.'" *Environmental Politics* 25 (4): 593–612. https://doi.org/10.1080/09644016.2016.1153771.

Boudet, Hilary, Christopher Clarke, Dylan Bugden, Edward Maibach, Connie Roser-Renouf, and Anthony Leiserowitz. 2014. "'Fracking' Controversy and Communication: Using National Survey Data to Understand Public Perceptions of Hydraulic Fracturing." *Energy Policy* 65 (February): 57–67. https://doi.org/10.1016/j.enpol.2013.10.017.

Boudet, Hilary, Chad M. Zanocco, Peter D. Howe, and Christopher E. Clarke. 2018. "The Effect of Geographic Proximity to Unconventional Oil and Gas Development on Public Support for Hydraulic Fracturing." *Risk Analysis: An Official Publication of the Society for Risk Analysis*. April. https://doi.org/10.1111/risa.12989.

Boushey, Graeme. 2010. *Policy Diffusion Dynamics in America*. Cambridge: Cambridge University Press.

Boyd, Brendan. 2017. "Working Together on Climate Change: Policy Transfer and Convergence in Four Canadian Provinces." *Publius: The Journal of Federalism* 47 (4): 546–71. https://doi.org/10.1093/publius/pjx033.

Boyd, Brendan, and Andrea Olive, eds. 2021. *Provincial Policy Laboratories: Policy Diffusion and Transfer in Canada's Federal System*. Toronto: University of Toronto Press.

Bradshaw, Michael, Patrick Devine-Wright, Darrick Evensen, Owen King, Abigail Martin, Stacia Ryder, Damien Short, Benjamin K. Sovacool, Paul Stretesky, Anna Szolucha, and Laurence Williams. 2022. "'We're Going All Out for Shale:' Explaining Shale Gas Energy Policy Failure in the United Kingdom." *Energy Policy* 168 (September): 113132. https://doi.org/10.1016/j.enpol.2022.113132.

Brasier, Kathryn J., Diane. K McLaughlin, Danielle Rhubart, Richard C. Stedman, Matthew R. Filteau, and Jeffrey Jacquet. 2013. "Risk Perceptions of Natural Gas Development in the Marcellus Shale." *Environmental Practice* 15 (2): 108–22. https://doi.org/10.1017/S1466046613000021

Bratt, Duane. 2020. "Remarkable Signs of Federal-Provincial Unity on Small Nuclear Reactors." *Policy Options* (blog). 30 March. https://policyoptions .irpp.org/magazines/march-2020/remarkable-signs-of-federal-provincial -unity-on-small-nuclear-reactors/.

– 2021. "Energy-Environment Federalism in Canada: Finding a Path for the Future." Ottawa, ON: Positive Energy, University of Ottawa. https://ruor .uottawa.ca/bitstream/10393/42493/1/Bratt%202021.pdf.

Brauers, Hanna. 2022. "Natural Gas as a Barrier to Sustainability Transitions? A Systematic Mapping of the Risks and Challenges." *Energy Research and Social Science* 89 (July): 102538. https://doi.org/10.1016/j.erss.2022.102538.

Brauers, Hanna, Isabell Braunger, and Jessica Jewell. 2021. "Liquefied Natural Gas Expansion Plans in Germany: The Risk of Gas Lock-In under Energy Transitions." *Energy Research & Social Science* 76 (June): 102059. https://doi.org/10.1016/j.erss.2021.102059.

Braul, Wally. 2011. "The Changing Regulatory Scheme in Northeast British Columbia" *Alberta Law Review* 49: 369. https://doi.org/10.29173/alr121

Breakwell, Glynis M. 2007. *The Psychology of Risk*. Cambridge: Cambridge University Press.

British Columbia New Democratic Party. 2013. "Change for the Better." Vancouver, B.C., Canada. https://www.poltext.org/sites/poltext.org/files/plateformesV2/Colombie-Britannique/BC_PL_2013_NDP_en.pdf.

Brown, Cassarah, Christopher Borick, Christopher Gore, Sarah Banas Mills, and Barry G. Rabe. 2014. "Shale Gas and Hydraulic Fracturing in the Great Lakes Region: Current Issues and Public Opinion." Ann Arbor, MI: Center for Local, State, and Urban Policy.

Bullock, Justin B., and Arnold Vedlitz. 2017. "Emphasis Framing and the Role of Perceived Knowledge: A Survey Experiment." *Review of Policy Research* 34 (4): 485–503. https://doi.org/10.1111/ropr.12231.

Bundale, Brett. 2011a. "Firm 'Hopeful' Vandals Won't Hit." *Halifax Chronicle Herald*, 6 October. http://thechronicleherald.ca/business/19184-firm-hopeful-vandals-won%E2%80%99t-hit.

– 2011b. "Waste Water Issue Mires Energy Firm." *Halifax Chronicle Herald*, 7 October. http://thechronicleherald.ca/business/19633-waste-water-issue-mires-energy-firm.

– 2011c. "N.S. Oil and Gas Firms Dismiss Fracking Report." *Halifax Chronicle Herald*, 9 December. http://thechronicleherald.ca/business/41175-ns-oil-and-gas-firms-dismiss-fracking-report.

– 2012a. "Corridor to Begin Exploration on Anticosti Island." *Halifax Chronicle Herald*, September. http://thechronicleherald.ca/business/132670-corridor-to-begin-exploration-on-anticosti-island.

– 2012b. "Fracking Cleanup Target Missed." *Halifax Chronicle Herald*, 29 November. http://thechronicleherald.ca/business/204208-fracking-cleanup-target-missed.

Buschmann, Pia, and Angela Oels. 2019. "The Overlooked Role of Discourse in Breaking Carbon Lock-In: The Case of the German Energy Transition." *WIREs Climate Change* 10 (3): e574. https://doi.org/10.1002/wcc.574.

Cairney, Paul. 2013. "Standing on the Shoulders of Giants: How Do We Combine the Insights of Multiple Theories in Public Policy Studies?" *Policy Studies Journal* 41 (1): 1–21. https://doi.org/10.1111/psj.12000.

Cairney, Paul, Kathryn Oliver, and Adam Wellstead. 2016. "To Bridge the Divide between Evidence and Policy: Reduce Ambiguity as Much as

Uncertainty." *Public Administration Review* 76 (3): 399–402. https://doi.org /10.1111/puar.12555.

Cairney, Paul, and Christopher M. Weible. 2017. "The New Policy Sciences: Combining the Cognitive Science of Choice, Multiple Theories of Context, and Basic and Applied Analysis." *Policy Sciences* 50 (4): 619–27. https://doi .org/10.1007/s11077-017-9304-2.

Calgary Herald. 2012. "NDP Pushes for Fracking Safety Review." 18 January. Sec. E. 4.

Campbell, Francis. 2014. "Fracking Drilled at Meeting." *Halifax Chronicle Herald*, 22 July. http://thechronicleherald.ca/novascotia/1224664-fracking -drilled-at-meeting.

Campbell, Karen, and Matt Horne. 2011. "Shale Gas in British Columbia: Risks to B.C.'s Water Resources." Drayton Valley, Alberta: The Pembina Institute.

Canada, Parliament. House of Commons. 2013. "Standing Committee on Natural Resources Minutes of Proceedings. 1st Sess., 41st Parliament, Meeting No. 066 2013." Government of Canada. http://www.parl.gc.ca /HousePublications/Publication.aspx?DocId=5979593&Language=E&Mo de=1.

Canadian Association of Physicians for the Environment (CAPE). 2020. "Fractures in the Bridge: Unconventional (Fracked) Natural Gas, Climate Change, and Human Health." Prepared by Ronald Macfarlane, and Kim Perrotta. January. https://cape.ca/wp-content/uploads/2020/01/CAPE -Fracking-Report-EN.pdf.

Canadian Association of Petroleum Producers (CAPP). 2011. "CAPP Members Establish New Guiding Principles for Hydraulic Fracturing – Canadian Association of Petroleum Producers." 8 September. https://web.archive .org/web/20120418165823/http://www.capp.ca/aboutUs/mediaCentre /NewsReleases/Pages/GuidingPrinciplesforHydraulicFracturing.aspx.

– 2014. "Canadian Marketed Gas Production 1947–2009." http://statshb .capp.ca/SHB/Sheet.asp?SectionID=3&SheetID=269.

– 2017. "Natural Gas Development." Canadian Association of Petroleum Producers. 2017. https://web.archive.org/web/20150405024125/http:// www.capp.ca/canadian-oil-and-natural-gas/natural-gas/natural-gas -development.

– 2022. "Petroleum Industry Statistics | The Statistical Handbook | CAPP." *CAPP | A Unified Voice for Canada's Upstream Oil and Gas Industry* (blog). 2022. https://www.capp.ca/resources/statistics/.

Canadian Election Survey. 2008. "Canadian Election Study 2008." www.odesi .ca.

Canadian Press. 2012a. "NS Puts Hold on Fracking for Now." *Halifax Chronicle Herald*, 16 April.http://thechronicleherald.ca/business/87357-ns-puts-hold -on-fracking-for-now.

– 2012b. "Cleanup of N.S. Fracking Waste Site Delayed." *Halifax Chronicle Herald*, 2 October. http://thechronicleherald.ca/business/142697-cleanup-of-ns-fracking-waste-site-delayed.

– 2014. "B.C. Fracking Challenged by Environmental Groups in Court." *Vancouver Sun*, 17 March. http://www.cbc.ca/news/canada/british-columbia/b-c-fracking-challenged-by-environmental-groups-in-court-1.2575719.

Canadian Centre for Energy Information (CCEI). 2007. "Canada's Evolving Offshore Oil and Gas Industry: Energy Today and Tommorow." Calgary: Canadian Centre for Energy Information. http://www.centreforenergy.com/Shopping/uploads/111.pdf.

Conservation Council of New Brunswick (CCNB). 2010. "Fracking for Shale Gas in New Brunswick: What You Need to Know." Fredericton, NB: Conservation Council of NB.

Cann, Heather W., and Leigh Raymond. 2018. "Does Climate Denialism Still Matter? The Prevalence of Alternative Frames in Opposition to Climate Policy." *Environmental Politics* 27 (3): 433–54. https://doi.org/10.1080/09644016.2018.1439353.

Cantoni, Roberto, Claudia Foltyn, Reiner Keller, and Matthias S. Klaes. 2023. "Introduction: What Is Fracking a Case Of?: Theoretical Lessons from European Case Studies." *Nature and Culture* 18 (1): 1–19. https://doi.org/10.3167/nc.2023.180101.

Carroll, Myles, Eleanor Stephenson, and Karena Shaw. 2012. "BC Political Economy and the Challenge of Shale Gas: Negotiating a Post-Staples Trajectory." *Canadian Political Science Review* 5 (2): 165–76. https://doi.org/10.24124/c677/2011324.

Carter, Angela V. 2020. *Fossilized : Environmental Policy in Canada's Petro-Provinces*. Vancouver: UBC Press.

Carter, Angela V., and Emily M. Eaton. 2016. "Subnational Responses to Fracking in Canada: Explaining Saskatchewan's 'Wild West' Regulatory Approach." *Review of Policy Research* 33 (4): 393–419. https://doi.org/10.1111/ropr.12179.

Carter, Angela V., Gail S. Fraser, and Anna Zalik. 2017. "Environmental Policy Convergence in Canada's Fossil Fuel Provinces? Regulatory Streamlining, Impediments, and Drift." *Canadian Public Policy* 43 (1): 61–76. https://doi.org/10.3138/cpp.2016-041

Carter, Angela V., and Leah M. Fusco. 2017. "Western Newfoundland's Anti-Fracking Campaign: Exploring the Rise of Unexpected Community Mobilization." *Journal of Rural and Community Development* 12 (1). http://journals.brandonu.ca/jrcd/article/view/1356.

Carter, Angela V., and Janetta McKenzie. 2020. "Amplifying 'Keep It in the Ground' First-Movers: Toward a Comparative Framework." *Society &*

Natural Resources 33 (11): 1339–58. https://doi.org/10.1080/08941920.2020
.1772924.

Cathles, Lawrence M., Larry Brown, Milton Taam, and Andrew Hunter.
2012. "A Commentary on 'The Greenhouse-Gas Footprint of Natural
Gas in Shale Formations' by R.W. Howarth, R. Santoro, and Anthony
Ingraffea." *Climatic Change* 113 (2): 525–35. https://doi.org/10.1007
/s10584-011-0333-0.

Cattaneo, Claudia. 2018. "LNG's Unlikely Saviour? B.C.'s NDP Premier Turns
Cheerleader in Asia Trip." *Financial Post*, 19 January. http://business
.financialpost.com/commodities/energy/lngs-unlikely-saviour-b-c-s-ndp
-premier-turns-cheerleader-in-asia-trip.

CBC News. 2009a. "CBC.ca – Nova Scotia Votes – Districts & Candidates." 9
June. https://web.archive.org/web/20091129040310/http://www.cbc.ca
/canada/nsvotes2009/ridings/.

– 2009b. "Quebec, N.B. Strike $4.8B Deal for NB Power." 29 October. https://
www.cbc.ca/news/canada/new-brunswick/quebec-n-b-strike-4-8b-deal
-for-nb-power-1.787566.

– 2010. "CBC.ca – New Brunswick Votes – Interactive Map." 27 September.
https://web.archive.org/web/20101129035358/http://www.cbc.ca
/canada/nbvotes2010/map/2010/.

– 2011. "Natural Gas Wells Fail to Meet Expectations – New Brunswick."
31 May. http://www.cbc.ca/news/canada/new-brunswick/natural-gas
-wells-fail-to-meet-expectations-1.1070793.

– 2013a. "Peace River North Election Results 2013." http://www.cbc.ca
/news2/canada/bcvotes2013/#/45.

– 2013b. "N.B. Academic Louis LaPierre's Credentials Face Scrutiny – New
Brunswick – CBC News." 6 September. http://www.cbc.ca/news/canada
/new-brunswick/n-b-academic-louis-lapierre-s-credentials-face-scrutiny
-1.1701555.

– 2013c. "Home – Nova Scotia Votes 2013 – CBC News." 9 October. https://
web.archive.org/web/20131231130440/http://www.cbc.ca/elections
/nsvotes2013/.

– 2014. "Shale Gas Moratorium Details Unveiled by Brian Gallant."
18 December. http://www.cbc.ca/1.2877440.

– 2015. "Yukon Government Says Yes to Fracking in Liard Basin." 11 April.
http://www.cbc.ca/news/canada/north/yukon-government-says-yes-to
-fracking-in-liard-basin-1.3028947.

Centner, Terence J., and Laura Kathryn O'Connell. 2014. "Unfinished Business
in the Regulation of Shale Gas Production in the United States." *Science of
The Total Environment* 476/477 (April): 359–67. https://doi.org/10.1016/j
.scitotenv.2013.12.112. Medline:24476976

Chailleux, Sébastien. 2020. "Strategic Ignorance and Politics of Time: How Expert Knowledge Framed Shale Gas Policies." *Critical Policy Studies* 14 (2): 174–92. https://doi.org/10.1080/19460171.2018.1563556.

Checkel, Jeffrey T. 2006. "Tracing Causal Mechanisms." *International Studies Review* 8 (2): 362–70. https://doi.org/10.1111/j.1468-2486.2006.00598_2.x.

Chen, Sibo. 2020a. "Debating Extractivism: Stakeholder Communications in British Columbia's Liquefied Natural Gas Controversy." *SAGE Open* 10 (4): 2158244020983007. https://doi.org/10.1177/2158244020983007.

– 2020b. "A Bridge to Where? Tracing the Bridge Fuel Metaphor in the Canadian Media Sphere." *Frontiers in Communication* 5 (December): 586711. https://doi.org/10.3389/fcomm.2020.586711.

Chilibeck, John. 2012. "Gov't Withholds Report on Fracking; Government Departments Need to Comment on Health Impact Study First, Says Minister." *Moncton Times and Transcript*. 4 October. Sec. Main.

– 2013. "Expert Calls for More Shale Gas Studies." *Moncton Times and Transcript*. 22 August. Sec. A.

Christopherson, Susan, Clay Frickey, and Ned Rightor. 2013. "A Vote of 'No Confidence' Why Local Governments Take Action in Response to Shale Gas Development." CRP Working Papers. Ithaca, NY: City and Regional Planning, Cornell University. https://web.archive.org/web/20140118223206/http://www.greenchoices.cornell.edu/downloads/development/shale/Vote_of_No_Confidence_WP.pdf.

Clancy, Peter. 2013. "Politics and Social Licence in New Brunswick's Shale Gas Sector." Presented at the annual conference of the Canadian Political Science Association. Victoria, BC.

Clark, Lisa F. 2013. "Framing the Uncertainty of Risk: Models of Governance for Genetically Modified Foods." *Science and Public Policy* (February): sct001. https://doi.org/10.1093/scipol/sct001.

Clarke, Christopher E., Dylan Bugden, P. Sol Hart, Richard C. Stedman, Jeffrey B. Jacquet, Darrick T.N. Evensen, and Hilary Boudet. 2016. "How Geographic Distance and Political Ideology Interact to Influence Public Perception of Unconventional Oil / Natural Gas Development." *Energy Policy* 97 (October): 301–9. https://doi.org/10.1016/j.enpol.2016.07.032.

Clarke, Christopher E., Philip S. Hart, Jonathon P. Schuldt, Darrick T. N. Evensen, Hilary Boudet, Jeffrey B. Jacquet, and Richard C. Stedman. 2015. "Public Opinion on Energy Development: The Interplay of Issue Framing, Top-of-Mind Associations, and Political Ideology." *Energy Policy* 81 (June): 131–40. https://doi.org/10.1016/j.enpol.2015.02.019.

Cleary, Eilish. 2012. "Chief Medical Officer of Health's Recommendations Concerning Shale Gas Development in New Brunswick." Fredericton: Office of the Chief Medical Officer of Health (OCMOH) New Brunswick Department of Health. http://www2.gnb.ca/content/dam/gnb

/Departments/h-s/pdf/en/HealthyEnvironments/Recommendations
_ShaleGasDevelopment.pdf.

Cleland, Michael, Stephen Bird, Stewart Fast, Shafak Sajid, and Louis Simard.
2016. "A Matter of Trust: The Role of Communities in Energy Decision-
Making." Ottawa: Canada West Foundation and the University of Ottawa.
http://www.uottawa.ca.positive-energy/files/secondversion_mattertrust
_report_24nov2016-1_web.pdf.

Cleland, Michael, and Monica Gattinger. 2017. "System Under Stress:
Energy Decision-Making in Canada and the Need for Informed
Reform." Ottawa: University of Ottawa. http://www.uottawa
.ca.positive-energy/files/2_positive_energy-system_under_stress-
cleland_and_gattinger.pdf.

Cobb, Roger W., and Charles D. Elder. 1971. "The Politics of Agenda-Building:
An Alternative Perspective for Modern Democratic Theory." *The Journal of
Politics* 33 (4): 892. https://doi.org/10.2307/2128415.

Coleman, Rich. 2013. "Development of LNG Promises Transformative Wealth
for the Province." *Vancouver Sun*, 22 July, sec. Issues & Ideas.

confidential interviewee. 2014a. Personal Interview. 10 November. Fredericton,
New Brunswick.

– 2014b. Personal Interview. 12 November. Fredericton, New Brunswick.

– 2014c. Personal Interview. 14 November. Fredericton New Brunswick.

– 2014d. Personal Interview. 17 November. Fredericton, New Brunswick.

– 2014e. Personal Interview. 19 November. Fredericton, New Brunswick.

– 2014f. Personal Interview. 19 November. Fredericton, New Brunswick.

– 2014g. Personal Interview. 24 November. Halifax, Nova Scotia.

– 2014h. Personal Interview. 27 November. Halifax, Nova Scotia.

– 2014i. Personal Interview. 28 November. Halifax, Nova Scotia.

– 2014j. Personal Interview. 4 December. St. John's, Newfoundland.

– 2014k. Personal Interview. 17 December. By telephone.

– 2015a. Personal Interview. 8 July. Edmonton, Alberta.

– 2015b. Personal Interview. 9 July. Calgary, Alberta.

– 2015c. Personal Interview. 14 July. Victoria, British Columbia.

– 2015d. Personal Interview. 21 July. Vancouver, British Columbia.

– 2015e. Personal Interview. 25 August. By telephone.

– 2018. Personal Interview. 12 January. By telephone.

Cook, Jeffrey J. 2014. "Who's Regulating Who? Analyzing Fracking Policy in
Colorado, Wyoming, and Louisiana." *Environmental Practice* 16 (2): 102–12.
https://doi.org/10.1017/S1466046614000027.

Coon, David. 2014. Personal Interview. 13 November. Fredericton, New
Brunswick.

Cotton, Matthew, Imogen Rattle, and James Van Alstine. 2014. "Shale Gas
Policy in the United Kingdom: An Argumentative Discourse Analysis."

Energy Policy 73 (October): 427–38. https://doi.org/10.1016/j.enpol
.2014.05.031.

Council of Canadian Academies. 2014. "Environmental Impacts of Shale Gas
Extraction in Canada." Ottawa: The Expert Panel on Harnessing Science
and Technology to Understand the Environmental Impacts of Shale Gas
Extraction, Council of Canadian Academies. http://www.scienceadvice.ca
/uploads/eng/assessments%20and%20publications%20and%20news%20
releases/shale%20gas/shalegas_fullreporten.pdf.

Cousineau, Sophie, Bertrand Marotte, and Rheal Seguin. 2012. "Quebec Gas in
Peril as PQ Signals Ban." *Globe and Mail*, 21 September. Sec. B.

Crow, Deserai A., and Andrea Lawlor. 2016. "Media in the Policy Process:
Using Framing and Narratives to Understand Policy Influences." *Review of
Policy Research* 33 (5): 472–91. https://doi.org/10.1111/ropr.12187.

Cudrilla Resources. n.d. "About Cuadrilla – Cuadrilla Resources." Accessed
26 July 2023. https://cuadrillaresources.uk/about-cuadrilla/.

Culpepper, Pepper D. 2011. *Quiet Politics and Business Power: Corporate Control
in Europe and Japan*. New York: Cambridge University Press.

Davene, Jeffrey. 2013. "Activists Press Parties on Fracking in Nova Scotia."
Halifax Chronicle Herald, 20 September. http://thechronicleherald.ca
/novascotia/1155740-activists-press-parties-on-fracking-in-nova-scotia.

Davis, Charles. 2012. "The Politics of 'Fracking': Regulating Natural Gas
Drilling Practices in Colorado and Texas." *Review of Policy Research* 29 (2):
177–91. https://doi.org/10.1111/j.1541-1338.2011.00547.x.

Davis, Charles, and Jonathan M. Fisk. 2014. "Energy Abundance or
Environmental Worries? Analyzing Public Support for Fracking in the
United States." *Review of Policy Research* 31 (1): 1–16. https://doi.org
/10.1111/ropr.12048.

Davis, Charles, and Katherine Hoffer. 2012. "Federalizing Energy? Agenda
Change and the Politics of Fracking." *Policy Sciences* 45 (3): 221–41. https://
doi.org/10.1007/s11077-012-9156-8.

Delaney, Gordon. 2014. "Plenty of Opposition at Fracking Panel Meeting."
Halifax Chronicle Herald, 24 July. http://thechronicleherald.ca
/novascotia/1225185-plenty-of-opposition-at-fracking-panel-meeting.

Delborne, Jason A., Dresden Hasala, Aubrey Wigner, and Abby Kinchy. 2020.
"Dueling Metaphors, Fueling Futures: 'Bridge Fuel' Visions of Coal and
Natural Gas in the United States." *Energy Research and Social Science* 61
(March): 101350. https://doi.org/10.1016/j.erss.2019.101350.

Demont, John. 2012. "Game Changer." *Halifax Chronicle Herald*,
17 February. http://thechronicleherald.ca/heraldmagazinev1i1
/64083-game-changer.

– 2014. "There May Be Oil in Those Basins." *Halifax Chronicle Herald*, 9
February. http://thechronicleherald.ca/business/61130-there-may-be-oil
-those-basins.

Department of Energy and Mines. 2016. "Moratorium on Hydraulic Fracturing to Continue Indefinitely." Government of New Brunswick. https://www2 .gnb.ca/content/gnb/en/news/news_release.2016.05.0462.html.

Department of Environmental and Local Government, New Brunswick. 2014. "Shale Gas: Summary of Environmental Research. An Annotated Bibliography." Department of Environment and Local Government.

Devine-Wright, Patrick, and Yuko Howes. 2010. "Disruption to Place Attachment and the Protection of Restorative Environments: A Wind Energy Case Study." *Journal of Environmental Psychology, Identity, Place, and Environmental Behaviour* 30 (3): 271–80. https://doi.org/10.1016/j.jenvp .2010.01.008.

Dimitrov, Radoslav S. 2016. "The Paris Agreement on Climate Change: Behind Closed Doors." *Global Environmental Politics* 16 (3): 1–11. https://doi.org /10.1162/GLEP_a_00361.

Dion, J., A. Kanduth, J. Moorhouse, and D. Beugin. 2021. "Canada's Net Zero Future: Finding Our Way in the Global Transition." Canadian Institute for Climate Choices. https://climatechoices.ca/reports/canadas-net-zero-future/.

Dobbin, Frank, Beth Simmons, and Geoffrey Garrett. 2007. "The Global Diffusion of Public Policies: Social Construction, Coercion, Competition, or Learning?" *Annual Review of Sociology* 33 (1): 449–72. https://doi.org /10.1146/annurev.soc.33.090106.142507.

Doberstein, Carey. 2016. *Building a Collaborative Advantage: Network Governance and Homelessness Policy-Making in Canada.* Vancouver: UBC Press.

– 2017. "The Credibility Chasm in Policy Research from Academics, Think Tanks, and Advocacy Organizations." *Canadian Public Policy* 43 (4): 363–75. https://doi.org/10.3138/cpp.2016-067.

Doberstein, Carey, and Heather Millar. 2014. "Balancing a House of Cards: Throughput Legitimacy in Canadian Governance Networks." *Canadian Journal of Political Scienc* 47 (June): 259–80. https://doi.org/10.1017 /S0008423914000420.

Doern, G. Bruce. 2005. "Canadian Energy Policy and the Struggle for Sustainable Development: Political-Economic Context." In *Canadian Energy Policy and the Struggle for Sustainable Development,* ed. G. Bruce Doern, 16–50. Toronto: University of Toronto Press.

Dolowitz, David, and David Marsh. 1996. "Who Learns What from Whom: A Review of the Policy Transfer Literature." *Political Studies* 44 (2): 343–57. https://doi.org/10.1111/j.1467-9248.1996.tb00334.x.

– 2002. "Learning from Abroad: The Role of Policy Transfer in Contemporary Policy-Making." *Governance* 13 (1): 5–23. https://doi.org/10.1111/0952 -1895.00121.

Douglas, Mary, and Aaron Wildavsky. 1983. *Risk and Culture: An Essay on the Selection of Technological and Environmental Dangers.* Oakland: University of California Press.

Druckman, James N. 2001a. "On the Limits of Framing Effects: Who Can Frame?" *The Journal of Politics* 63 (4): 1041–66. https://doi.org/10.1111/0022-3816.00100.

– 2001b. "The Implications of Framing Effects for Citizen Competence." *Political Behavior* 23 (3): 225–56. https://doi.org/10.1023/A:1015006907312.

Druckman, James N., and Arthur Lupia. 2000. "Preference Formation." *Annual Review of Political Science* 3 (1): 1–24. https://doi.org/10.1146/annurev.polisci.3.1.1.

Dryzek, John S. 2021. *The Politics of the Earth*. 4th ed. New York: Oxford University Press.

Duffy, Robert. 2023. "'The Cleanest Oil Fields in the World': Narratives Encouraging Oil Development in California." *Energy Research and Social Science* 97 (March): 102998. https://doi.org/10.1016/j.erss.2023.102998.

Dufour, Pascale, Laurence Bherer, and Christine Rothmayr. 2011. "Luttes Contre l'exploitation Des Gaz de Schistes Au Québec : Quand Un Enjeu Environnemental Brasse Les Cartes Du Jeu Politique." *Mouvements* (December). http://mouvements.info/luttes-contre-lexploitation-des-gaz-de-schistes-au-quebec-quand-un-enjeu-environnemental-brasse-les-cartes-du-jeu-politique/.

Dundon, Leah A., Mark Abkowitz, and Janey Camp. 2015. "The Real Value of FracFocus as a Regulatory Tool: A National Survey of State Regulators." *Energy Policy* 87 (C): 496–504. https://doi.org/10.1016/j.enpol.2015.09.031

Dunlop, Claire A. 2017. "The Irony of Epistemic Learning: Epistemic Communities, Policy Learning and the Case of Europe's Hormones Saga." *Policy and Society* 36 (2): 215–32. https://doi.org/10.1080/14494035.2017.1322260.

Dunlop, Claire A., and Claudio M. Radaelli. 2013. "Systematising Policy Learning: From Monolith to Dimensions." *Political Studies* 61 (3): 599–619. https://doi.org/10.1111/j.1467-9248.2012.00982.x.

– 2016. "Policy Learning in the Eurozone Crisis: Modes, Power and Functionality." *Policy Sciences* 49 (2): 107–24. https://doi.org/10.1007/s11077-015-9236-7. Medline:27471328

– 2017. "Learning in the Bath-Tub: The Micro and Macro Dimensions of the Causal Relationship between Learning and Policy Change." *Policy and Society* 36 (2): 304–19. https://doi.org/10.1080/14494035.2017.1321232.

– 2018a. "Does Policy Learning Meet the Standards of an Analytical Framework of the Policy Process?" *Policy Studies Journal* 46 (S1): S48–68. https://doi.org/10.1111/psj.12250. Medline:30034066

– 2018b. "The Lessons of Policy Learning: Types, Triggers, Hindrances and Pathologies." *Policy and Politics* 46 (2): 255–72. https://doi.org/10.1332/030557318X15230059735521.

Eaton, Emily, and Abby Kinchy. 2016. "Quiet Voices in the Fracking Debate: Ambivalence, Nonmobilization, and Individual Action in Two Extractive

Communities (Saskatchewan and Pennsylvania)." *Energy Research and Social Science* 20 (October): 22–30. https://doi.org/10.1016/j.erss.2016.05.005.

Ecojustice. 2013. "Legal Backgrounder: Bill 2 Responsible Energy Development Act." https://www.ecojustice.ca/wp-content/uploads/2015/03/REDA -backgrounder-May-2013.pdf.

EKOS Research Associates Inc. 2016. "Canadian Attitudes towards Energy and Pipelines: Survey Findings." Ottawa, ON: EKOS. https://web.archive.org /web/20160624162013/http://www.ekospolitics.com/wp-content /uploads/full_report_march_17_2016.pdf.

Emery, Herb. 2020. "Economic Survival Means Learning from the Past." *Atlantic Institute for Policy Research* (blog). 26 May. https://blogs.unb.ca /aipr/2020/05/lessons-from-the-past.php.

– 2021. "No CANDU? Why Refitting Point Lepreau Wasn't So Crazy, after All." *Telegraph Journal*. https://tj.news/telegraph-journal/101683015.

Energy and Mines Minister's Conference. 2013a. "Annex B: Shale Resources Compendium." Conference report, 1–43. Yellowknife, NWT: Government of Canada. http://www.nrcan.gc.ca/sites/www.nrcan .gc.ca/files/www/pdf/publications/emmc/AnnexB_Shale _Compemdium_e.pdf.

– 2013b. "Responsible Shale Development Enhancing the Knowledge Base on Shale Oil and Gas in Canada." Conference report, 1–43. Yellowknife, NWT: Government of Canada. http://www.nrcan.gc.ca/sites/www.nrcan.gc.ca /files/www/pdf/publications/emmc/Shale_Resources_e.pdf.

Energy Resources Conservation Board (ERCB). 2011. "Unconventional Gas Regulatory Framework – Jurisdictional Review." Calgary: Energy Resources Conservation Board. http://www.aer.ca/documents/reports /r2011-A.pdf.

– 2012a. "A Discussion Paper: Regulating Unconventional Oil and Gas in Alberta." Calgary: Energy Resources Conservation Board. http://www.aer .ca/documents/projects/URF/URF_DiscussionPaper_20121217.pdf.

– 2012b. "News Release 2012-12-17 (NR2012-13) ERCB Seeking Feedback on Regulatory Approach for Unconventional Development." Calgary: Energy Resources Conservation Board. https://web.archive.org/web /20131007103857/https://www.aer.ca/documents/news-releases/NR2012 -13.pdf.

– 2012c. "Bulletin 2012–25: Amendments to Directive 059: Well Drilling and Completion Data Filing Requirements in Support of Disclosure of Hydraulic Fracturing Fluid Information." Calgary: Energy Resources Conservation Board. https://web.archive.org/web/20130926082629 /http://www.aer.ca/documents/bulletins/Bulletin-2012-25.pdf.

Environment and Climate Change Canada (ECCC). 2016. "Pan-Canadian Framework on Clean Growth and Climate Change : Canada's Plan to

Address Climate Change and Grow the Economy." Ottawa. https://
publications.gc.ca/collections/collection_2017/eccc/En4-294-2016-eng.pdf.

Environment and Local Government. 2012. "A Guide to Environmental
Impact Assessment in New Brunswick." Fredericton: Government of New
Brunswick. http://www2.gnb.ca/content/dam/gnb/Departments/env
/pdf/EIA-EIE/GuideEnvironmentalImpactAssessment.pdf.

Ernst and Young. 2013. "Global LNG: Will New Demand and New Supply
Mean New Pricing?" London: EYGM Limited. http://www.ey.com
/Publication/vwLUAssets/Global_LNG_New_pricing_ahead/$FILE/Global
_LNG_New_pricing_ahead_DW0240.pdf.

Ernst & Young Global (EY). 2015a. "Alberta's Oil and Gas Sector Regulatory
Paradigm Shift: Challenges and Opportunities." http://www.ey.com
/Publication/vwLUAssets/EY-Alberta-oil-gas-regulatory-paradigm-shift
/$FILE/EY-Alberta-oil-gas-regulatory-paradigm-shift.pdf.

– 2015b. "Review of British Columbia's Hydraulic Fracturing Regulatory
Framework." Victoria: BC Oil and Gas Commission. https://www.bc-er
.ca/files/reports/Seismicity-and-Fracturing/bcogc-hf-regulatory-review
-appendix-c.pdf.

Erskine, Bruce. 2014. "Nova Scotia to Ban Fracking." *Halifax Chronicle Herald*, 3
September. http://thechronicleherald.ca/business/1233818-nova-scotia-to
-ban-fracking.

Evensen, Darrick. 2018. "Review of Shale Gas Social Science in the United
Kingdom, 2013–2018." *The Extractive Industries and Society* 5 (4): 691–8.
https://doi.org/10.1016/j.exis.2018.09.005.

Evensen, Darrick, Jeffrey B. Jacquet, Christopher E. Clarke, and Richard C.
Stedman. 2014. "What's the 'Fracking' Problem? One Word Can't Say It
All." *The Extractive Industries and Society* 1 (2): 130–6. https://doi.org
/10.1016/j.exis.2014.06.004.

Evensen, Darrick, and Rich Stedman. 2017. "Beliefs about Impacts Matter
Little for Attitudes on Shale Gas Development." *Energy Policy* 109
(Supplement C): 10–21. https://doi.org/10.1016/j.enpol.2017.06.053.

Falkner, Robert, and Nico Jaspers. 2012. "Regulating Nanotechnologies: Risk,
Uncertainty and the Global Governance Gap." *Global Environmental Politics*
12 (1): 30–55. https://doi.org/10.1162/GLEP_a_00096.

Fast, Stewart. 2016a. "A Matter of Trust: The Role of Communities in Energy
Decision-Making – Shale Gas Exploration Case Study: Kent County and
Elsipogtog First Nation New Brunswick." Ottawa: Canada West Foundation
and the University of Ottawa. https://www.uottawa.ca.positive-energy
/files/nrp_mattertrust_casestudy_kentcounty_24nov2016.pdf.

– 2016b. "Assessing Public Participation Tools during Wind Energy Siting."
Journal of Environmental Studies and Sciences (December): 1–8. https://doi.org
/10.1007/s13412-016-0419-0.

Fisk, Jonathan M. 2013. "The Right to Know? State Politics of Fracking Disclosure." *Review of Policy Research* 30 (4): 345–65. https://doi.org/10.1111/ropr.12025.

Fisk, Jonathan M., Yunmi Park, and Zachary Mahafza. 2017. "'Fractivism' in the City: Assessing Defiance at the Neighborhood Level." *State and Local Government Review* 49 (2): 105–16. https://doi.org/10.1177/0160323X17720712.

Foss, Michelle. 2007. "United States Natural Gas Prices to 2015." Oxford: The Oxford Institute for Energy Studies. http://www.oxfordenergy.org/2007/02/united-states-natural-gas-prices-to-2015/.

– 2011. "The Outlook for US Gas Prices in 2020: Henry Hub at $3 or $10?" Oxford: The Oxford Institute for Energy Studies. http://www.oxfordenergy.org/wpcms/wp-content/uploads/2011/12/NG_58.pdf.

Foster, James. 2010. "Turtle Creek Drilling Panned; Moncton City Council Wants Province to Ban Oil, Gas Exploration in Watershed." *Moncton Times and Transcript*, 16 November.

– 2011a. "U.S. Gas Rules Explored; N.B. Delegation Checks Out Natural Gas Regulations Stateside." *Moncton Times and Transcript*, 25 January, sec. International.

– 2011b. "We'll Do It Right: DNR Minister; Northrup Says Shale Gas to Proceed Safely or Not at All." *Moncton Times and Transcript*, 14 March, sec. Main.

Fowlie, Jonathan. 2013. "Liberal Candidate Calls for Scientific Review of Hydraulic Fracking; Clark Does Not Support Bernier's Position." *Vancouver Sun*, 3 May. Sec. Westcoast News.

Fox, Josh, dir. 2010. *Gasland*. HBO.

Fraser Basin Council. 2012. "Identifying Health Concerns Relating to Oil and Gas Development in Northeastern BC: Human Health Risk Assessment – Phase 1 Report." Vancouver: Fraser Basin Council. http://www.health.gov.bc.ca/library/publications/year/2012/Identifying-health-concerns-HHRA-Phase1-Report.pdf.

Gagnon, G.A., W. Krkosek, L. Anderson, E. McBean, M. Mohseni, M. Bazri, and I. Mauro. 2015. "Impacts of Hydraulic Fracturing on Water Quality: A Review of Literature, Regulatory Frameworks and an Analysis of Information Gaps." *Environmental Reviews* 24 (2): 122–31. https://doi.org/10.1139/er-2015-0043.

Gallant, Jacques. 2011. "N.B. Moving 'at Right Speed' on Shale Gas: Premier Alward Says Moratorium on Shale Gas Exploration Not Necessary, Even in Light of Recent Protests." *Moncton Times and Transcript*, 15 August.

Garvie, Kathryn H., Lana Lowe, and Karena Shaw. 2014. "Shale Gas Development in Fort Nelson First Nation Territory: Potential Regional

Impacts of the LNG Boom." *BC Studies: The British Columbian Quarterly*, no. 184 (Winter): 45–72. https://doi.org/10.14288/bcs.v0i184.184887.

Garvie, Kathryn H., and Karena Shaw. 2014. "Oil and Gas Consultation and Shale Gas Development in British Columbia." *BC Studies: The British Columbian Quarterly*, no. 184 (Winter): 73–102.

– 2016. "Shale Gas Development and Community Response: Perspectives from Treaty 8 Territory, British Columbia." *Local Environment* 21 (8): 1009–28. https://doi.org/10.1080/13549839.2015.1063043.

Gastil, John, Don Braman, Dan Kahan, and Paul Slovic. 2011. "The Cultural Orientation of Mass Political Opinion." *PS: Political Science and Politics* 44 (4): 711–14. https://doi.org/10.1017/S1049096511001326.

Gattinger, Monica. 2012. "Canada–United States Energy Relations: Making a MESS of Energy Policy." *American Review of Canadian Studies* 42 (4): 460–73. https://doi.org/10.1080/02722011.2012.732331.

– 2015. "A National Energy Strategy for Canada: Golden Age or Golden Cage of Energy Federalism?" In *Canada: The State of the Federation 2012: Regions, Resources, and Resiliency*, ed. Loleen Berdahl, Andre Juneau, and Carolyn Hughes Tuohy, 39–70. Kingston: McGill Queens University Press.

Gattinger, Monica, and Rafael Aguirre. 2016. "The Shale Revolution and Canada-United States Energy Relations: Game Charger or déjà vu All Over Again?" In *International Political Economy*, ed. Greg Anderson and Christopher J. Kukucha, 409–35. Toronto: Oxford University Press.

Geny, Florence. 2010. "Can Unconventional Gas Be a Game Changer in European Gas Markets?" Oxford: The Oxford Institute for Energy Studies. http://www.oxfordenergy.org/2010/12/can-unconventional-gas-be-a -game-changer-in-european-gas-markets/.

Gerein, Keith. 2012. "MLAs Endure All-Nighter to Pass Energy Bill." *Calgary Herald*, 22 November. Sec. News.

Gerring, John. 2008. "The Mechanismic Worldview: Thinking inside the Box." *British Journal of Political Science* 38 (1): 161–79. https://doi.org/10.1017 /S0007123408000082.

Gilardi, Fabrizio. 2010. "Who Learns from What in Policy Diffusion Processes?" *American Journal of Political Science* 54 (3): 650–66. https://doi .org/10.1111/j.1540-5907.2010.00452.x.

Global News. 2019. "New Brunswick Indigenous Chiefs Left 'Blindsided' by Decision to Lift Fracking Moratorium," 5 June. https://globalnews.ca /news/5356115/indigenous-chiefs-issue-warning-gas-fracking/.

Goehner, Adam, and Matt Horne. 2014. "Wellhead to Waterline: Opportunities to Limit Greenhouse Gas Emissions from BC's Proposed LNG Industry." Calgary: Pembina Institute. http://www.pembina.org/reports/pi-wellhead -to-waterline-goehnerhorne-022014.pdf.

Goldthau, Andreas. 2018. *The Politics of Shale Gas in Eastern Europe*. Cambridge Studies in Comparative Public Policy. Cambridge: Cambridge University Press. https://doi.org/10.1017/9781316875018.

Goldthau, Andreas, and Benjamin K. Sovacool. 2016. "Energy Technology, Politics, and Interpretative Frames: Shale Gas Fracking in Eastern Europe." *Global Environmental Politics* 16 (4): 50–69. https://doi.org/10.1162/GLEP_a_00375.

Gorman, Michael. 2011. "Energy Expert Raises Risks of Fracking." *Halifax Chronicle Herald*, 4 December. http://thechronicleherald.ca/novascotia/39237-energy-expert-raises-risks-fracking.

– 2012. "Radioactivity Worries Hold Up Treatment of Fracking Wastewater." *Halifax Chronicle Herald*, 2 October. http://thechronicleherald.ca/novascotia/142764-radioactivity-worries-hold-up-treatment-of-fracking-wastewater.

– 2013a. "Colchester Approves Fracking Water Disposal in County Sewer System." *Halifax Chronicle Herald*, 27 March. http://thechronicleherald.ca/novascotia/1119954-colchester-approves-fracking-water-disposal-in-county-sewer-system.

– 2013b. "Residents Pour Cold Water on Proposal." *Halifax Chronicle Herald*, 8 April. http://thechronicleherald.ca/novascotia/1122290-residents-pour-cold-water-on-proposal.

– 2013c. "Fracking Waste Water Plans Rapped Again." *Halifax Chronicle Herald*, 7 May. http://thechronicleherald.ca/novascotia/1128132-fracking-waste-water-plans-rapped-again.

– 2013d. "CBU President to Lead Fracking Review." *Halifax Chronicle Herald*, 28 August. http://thechronicleherald.ca/novascotia/1150396-cbu-president-to-lead-fracking-review.

– 2013e. "Fracking Waste Targeted." *Halifax Chronicle Herald*, 3 December. http://thechronicleherald.ca/novascotia/1171407-fracking-waste-targeted.

– 2014a. "Aggressive Fracking Opponents May Be Hijacking Public Sessions: Younger." *Halifax Chronicle Herald*, 24 July. http://thechronicleherald.ca/novascotia/1225050-aggressive-fracking-opponents-may-be-hijacking-public-sessions-younger.

– 2014b. "Report: Nova Scotia Not Ready for Fracking." *Halifax Chronicle Herald*, 24 August. http://thechronicleherald.ca/novascotia/1232522-report-nova-scotia-not-ready-for-fracking.

Gormley, William T., Jr. 1986. "Regulatory Issue Networks in a Federal System." *Polity* 18 (4): 595–620. https://doi.org/10.2307/3234884.

Government of Alberta. 2020. "Natural Gas Vision and Strategy." Edmonton, Alberta. https://www.alberta.ca/natural-gas-vision-and-strategy.aspx.

Government of British Columbia. 2002. "Energy for Our Future: A Plan for BC."
Victoria, BC. https://web.archive.org/web/20120302033210/http://www
.bcenergyblog.com/uploads/file/2002%20BC%20Energy%20Plan.pdf.

– 2007a. "BC Energy Plan: Oil and Gas Policy Actions." Victoria, BC: Ministry
of Energy, Mines, and Petroleum Resources. https://web.archive.org/web
/20120210082946/http://www.energyplan.gov.bc.ca/PDF/BC_Energy
_Plan_Oil_and_Gas.pdf.

– 2007b. "The BC Energy Plan: A Vision for Clean Energy Leadership."
Victoria, BC: Ministry of Energy, Mines, and Petroleum Resources.
https://web.archive.org/web/20120203035826/http://www.energyplan
.gov.bc.ca/PDF/BC_Energy_Plan.pdf.

– 2012a. "British Columbia's Natural Gas Strategy: Fuelling BC's Economy for
the Next Decade and Beyond." Victoria, BC: Ministry of Energy and Mines.
https://web.archive.org/web/20130324070018/http://www.gov.bc.ca/ener
/popt/down/natural_gas_strategy.pdf.

– 2012b. "LNG: Liquefied Natural Gas A Strategy for B.C.'s Newest Industry."
Victoria, BC: Ministry of Energy and Mines. https://web.archive.org/web
/20130324070018/http://www.gov.bc.ca/ener/popt/down/natural_gas
_strategy.pdf.

– 2012c. "Canada's First Hydraulic Fracturing Registry Now Online." https://
news.gov.bc.ca/stories/canadas-first-hydraulic-fracturing-registry-now-online.

– 2016. "Climate Leadership Plan." Victoria, BC: Government of BC.
https://climate.gov.bc.ca/app/uploads/sites/13/2016/10/4030_CLP
_Booklet_web.pdf.

Government of Canada. 2021. "Canada's Enhanced Nationally Determined
Contribution." Backgrounders. 23 April. https://www.canada.ca/en
/environment-climate-change/news/2021/04/canadas-enhanced-nationally
-determined-contribution.html.

Government of Canada, National Energy Board. 2018. "NEB – ARCHIVED –
Western Regulators' Forum." 16 February. https://web.archive.org/web
/20170615032050/https://www.neb-one.gc.ca/bts/nws/whtnw/archive
/2014/2014-09-29-eng.html.

Government of New Brunswick. 2006. "Climate Change Action Plan 2007–2012."
Fredericton, NB: Department of the Environment. https://www2.gnb.ca
/content/dam/gnb/Departments/env/pdf/Climate-Climatiques/2007
-2012ClimateChangeActionPlan.pdf.

– 2009. "News Release: Minister to Promote New Brunswick Oil and Gas
Exploration." http://www2.gnb.ca/content/gnb/en/news/news
_release.2009.02.0105.html.

– 2011. "Make an Informed Decision: The Facts about Shale Gas." Fredericton.

– 2012. "Journals of the Legislative Assembly Second Session, Fifty-Seventh
Legislature 2011–2012: Journal 8 – Tuesday, December 6, 2011." Fredericton:

Government of New Brunswick. http://www.gnb.ca/legis/business
/currentsession/57/57-2/journals-e/08111206e.pdf.

– 2013a. "Responsible Environmental Management of Oil and Natural Gas
Activities in New Brunswick: Rules for Industry." Fredericton: Government
of New Brunswick. https://web.archive.org/web/20131126165104
/http://www2.gnb.ca/content/dam/gnb/Corporate/pdf/ShaleGas/en
/RulesforIndustry.pdf.

– 2013b. "The New Brunswick Oil and Natural Gas Blueprint." Fredericton:
Government of New Brunswick. https://web.archive.org/web
/20131126165104/http://www2.gnb.ca/content/dam/gnb/Corporate
/pdf/ShaleGas/en/RulesforIndustry.pdf.

– 2013c. "57th Legislature/Session 4, Reply to the Speech from the Throne/
Hansard/Proceedings." https://www.gnb.ca/legis/business/pastsessions
/57/57legislature-e.asp.

– n.d. "L'Explorateur ONG Map Viewer." Accessed 20 November 2014.
http://geonb.snb.ca/ong/.

Government of Newfoundland and Labrador. 2013. "Minister Provides
Position on Hydraulic Fracturing." Department of Natural Resources.
http://www.releases.gov.nl.ca/releases/2013/nr/1104n06.htm.

– 2016. "Provincial Government Receives Independent Hydraulic Fracturing
Panel Report." Department of Natural Resources. http://www.releases.gov
.nl.ca/releases/2016/nr/0531n06.aspx.

Government of Nova Scotia. 2007. *Environmental Goals and Sustainable Prosperity
Act*. http://nslegislature.ca/legc/bills/60th_1st/3rd_read/b146.htm.

– 2011a. "Province to Review Hydraulic Fracturing in Shale Gas Operations."
4 April. http://novascotia.ca/news/release/?id=20110404012.

– 2011b. "Final Scope: Review of Hydraulic Fracturing in Oil and Gas
Operations in Nova Scotia." Nova Scotia Department of Energy. https://
www.novascotia.ca/nse/pollutionprevention/docs/Consultation
.Fracturing.Scope.pdf.

– 2011c. "Frequently Asked Questions." Nova Scotia Department of Energy.
https://www.novascotia.ca/nse/pollutionprevention/docs/Consultation
.Hydraulic.Fracturing.Review.FAQ.pdf.

– 2011d. "Review of Hydraulic Fracturing in Nova Scotia." Nova Scotia
Department of Energy. https://www.novascotia.ca/nse
/pollutionprevention/docs/Consultation.Hydraulic.Fracturing
-What.We.Heard.pdf.

– 2012a. "Frequently Asked Questions Hydraulic Fracturing Review
Extension." http://www.novascotia.ca/nse/pollutionprevention/docs
/Consultation.FAQ.2012.Review.Extension.pdf.

– 2012b. "Province Extends Hydraulic Fracturing Review." 16 April.
http://novascotia.ca/news/release/?id=20120416004.

– 2013a. "Province Announces Independent Hydraulic Fracturing Review." 28 August. http://novascotia.ca/news/release/?id=20130828001.
– 2013b. "Assembly 62, Session 1 / Hansard / Proceedings / The Nova Scotia Legislature." https://web.archive.org/web/20150416192219/http:// nslegislature.ca/index.php/proceedings/hansard/C94/house_13dec12 /#HPage852.
– 2014a. "Hydraulic Fracturing Review | Pollution Prevention | Nova Scotia Environment." http://www.novascotia.ca/nse/pollutionprevention /consultation.hydraulic.fracturing.asp.
– 2014b. "Assembly 62, Session 2 / Hansard / Proceedings / The Nova Scotia Legislature." https://web.archive.org/web/20150414213519 /http://nslegislature.ca/index.php/proceedings/hansard/C96/house _14nov14/#HPage2576.
Government of Wales. 2018. "Licensing Powers on 'Fracking' Transferred to Wales." 1 October. https://www.gov.wales/licensing-powers-fracking -transferred-wales.
Graham, Nicolas, Shannon Daub, and Bill Carroll. 2017. "Mapping Political Influence: Political Donations and Lobbying by the Fossil Fuel Industry in BC." Vancouver: CCPA. https://www.policyalternatives.ca/sites/default /files/uploads/publications/BC%20Office/2017/03/ccpa-bc_mapping _influence_final.pdf.
Gravelle, Timothy B., and Erick Lachapelle. 2015. "Politics, Proximity and the Pipeline: Mapping Public Attitudes toward Keystone XL." *Energy Policy* 83 (August): 99–108. https://doi.org/10.1016/j.enpol.2015.04.004.
Grin, John, and Anne Loeber. 2006. "Theories of Policy Learning: Agency, Structure, and Change." In *Handbook of Public Policy Analysis: Theory, Politics, and Methods*, ed. Frank Fischer, Gerald J. Miller, and Mara S. Sidney, 201–19. Boca Raton, FL: CRC Press.
Haas, Peter. 1992. "Introduction: Epistemic Communities and International Policy Coordination." *International Organization* 46 (1): 1–35. https://doi.org /10.1017/S0020818300001442.
– 2004. "When Does Power Listen to Truth? A Constructivist Approach to the Policy Process." *Journal of European Public Policy* 11 (4): 569–92. https://doi .org/10.1080/1350176042000248034.
Hajer, Maarten A. 1993. "Discourse Coalitions and the Institutionalization of Practice: The Case of Acid Rain in Britain." In *The Argumentative Turn in Policy Analysis and Planning*, ed. Frank Fischer and John Forester, 43–76. Durham, NC: Duke University Press.
Halifax Chronicle Herald. 2014. "Amherst Walks Away from Waste Water Disposal Negotiations." 26 November. http://thechronicleherald.ca /novascotia/1253527-amherst-walks-away-from-waste-water-disposal -negotiations.

Halifax Chronicle Herald Editorial Board. 2014. "Fracking Ban Hastiness Backfires in Nova Scotia." 11 September. http://thechronicleherald.ca /editorials/1235702-editorial-fracking-ban-hastiness-backfires-in-nova -scotia.

Hall, Peter A. 1993. "Policy Paradigms, Social Learning, and the State: The Case of Economic Policymaking in Britain." *Comparative Politics* 25 (3): 275–96. https://doi.org/10.2307/422246.

– 2003. "Aligning Ontology and Methodology in Comparative Research." In *Comparative Historical Analysis in the Social Sciences*, ed. James Mahoney and Dietrich Rueschemeyer, 373–404. Cambridge: Cambridge University Press.

Harrison, Kathryn. 2012. "A Tale of Two Taxes: The Fate of Environmental Tax Reform in Canada." *Review of Policy Research* 29 (3): 383–407. https://doi.org /10.1111/j.1541-1338.2012.00565.x.

– 2013. "Federalism and Climate Policy Innovation: A Critical Reassessment." *Canadian Public Policy*, September. https://doi.org/10.3138/CPP.39 .Supplement2.S95.

– 2020. "Political Institutions and Supply-Side Climate Politics: Lessons from Coal Ports in Canada and the United States." *Global Environmental Politics* 20 (4): 51–72. https://doi.org/10.1162/glep_a_00579

– 2023. "Climate Governance and Federalism in Canada." In *Climate Governance and Federalism: A Forum of Federations Comparative Policy Analysis*, ed. Alan Fenna, Sébastien Jodoin, and Joana Setzer, 64–85. Cambridge: Cambridge University Press. https://doi.org/10.1017/9781009249676.

Harrison, Kathryn, and Guri Bang. 2022. "Supply-Side Climate Policies in Major Oil-Producing Countries: Norway's and Canada's Struggles to Align Climate Leadership with Fossil Fuel Extraction." *Global Environmental Politics* 22 (4): 129–50. https://doi.org/10.1162/glep_a_00682.

Harrison, Kathryn, and George Hoberg. 1994. *Risk, Science, and Politics: Regulating Toxic Substances in Canada and the United States*. Montreal: Carleton University Press.

Hartley, Sarah, and Grace Skogstad. 2005. "Regulating Genetically Modified Crops and Foods in Canada and the United Kingdom: Democratizing Risk Regulation." *Canadian Public Administration* 48 (3): 305–27. https://doi.org /10.1111/j.1754-7121.2005.tb00228.x.

Hay, Colin. 2010. "Ideas and the Construction of Interests." In *Ideas and Politics in Social Science Research*, ed. Daniel Béland and Robert Henry Cox, 65–82. New York: Oxford University Press.

Hayhurst, Ruth. 2023. "3 Onshore Oil and Gas Wells Drilled in 2022 – Official Data." *DRILL OR DROP?* (blog). 1 January. https://drillordrop.com /2023/01/01/3-onshore-oil-and-gas-wells-drilled-in-2022-official-data/.

Hays, Jake, and Seth B.C. Shonkoff. 2016. "Toward an Understanding of the Environmental and Public Health Impacts of Unconventional Natural Gas

Development: A Categorical Assessment of the Peer-Reviewed Scientific Literature, 2009–2015." *PLOS ONE* 11 (4): e0154164. https://doi.org/10.1371/journal.pone.0154164. Medline:27096432

Hazboun, Shawn Olson, and Hilary Schaffer Boudet. 2021. "Natural Gas – Friend or Foe of the Environment? Evaluating the Framing Contest over Natural Gas through a Public Opinion Survey in the Pacific Northwest." *Environmental Sociology* 7 (4), 368–81. https://doi.org/10.1080/23251042.2021.1904535.

Healing, Dan. 2013. "Investments in Duvernay Grow; $6B Has Been Spent on Shale Formation." *Calgary Herald*, 27 August. Sec. Business. https://www.pressreader.com/canada/calgary-herald/20130827/282376922239124.

Heclo, Hugh. 1974. *Modern Social Politics in Britain and Sweden: From Relief to Income Maintenance*. New Haven: Yale University Press. Reprint, Colchester, UK: ECPR Press, 2010.

Heerema, Dylan, and Maximilian Kniewasser. 2017. "Liquefied Natural Gas, Carbon Pollution and British Columbia in 2017." Pembina Institute. http://www.pembina.org/reports/lng-carbon-pollution-bc-2017.pdf.

Heffernan, Kevin. 2013. "Hydraulic Fracturing and Oil and Gas Development." Presented at the Calgary Chamber of Commerce Environment and Natural Resource Committee, Calgary, 11 April. http://www.csur.com/images/CSUG_presentations/2013/C_of_C_Calgary_April_13.pdf.

Heikkila, Tanya, and Andrea K. Gerlak. 2013. "Building a Conceptual Approach to Collective Learning: Lessons for Public Policy Scholars." *Policy Studies Journal* 41 (3): 484–512. https://doi.org/10.1111/psj.12026.

Heikkila, Tanya, Jonathan J. Pierce, Samuel Gallaher, Jennifer Kagan, Deserai A. Crow, and Christopher M. Weible. 2014. "Understanding a Period of Policy Change: The Case of Hydraulic Fracturing Disclosure Policy in Colorado." *Review of Policy Research* 31 (2): 65–87. https://doi.org/10.1111/ropr.12058.

Henton, Darcy. 2011. "Alberta Shuns U.S. Shale Gas Drilling Fears; Minister Says Our Rules Prevent Water Contamination." *Calgary Herald*, 14 May.

– 2012. "Energy Bill Seen as 'Train Wreck'; Province Defends Plan to Simplify Approvals." *Calgary Herald*, 5 November. Sec. News.

Hessing, Melody, Michael Howlett, and Tracy Summerville. 2005. *Canadian Natural Resource and Environmental Policy: Political Economy and Public Policy*. Vancouver: UBC Press.

Hoberg, George. 1991. "Sleeping with an Elephant: The American Influence on Canadian Environmental Regulation." *Journal of Public Policy* 11 (1): 107–31. https://doi.org/10.1017/S0143814X00004955.

– 1996. "Putting Ideas in Their Place: A Response to 'Learning and Change in the British Columbia Forest Policy Sector.'" *Canadian Journal of Political Science/Revue Canadienne de Science Politique* 29 (1): 135–44. https://doi.org/10.1017/S0008423900007277.

– 2013. "The Battle over Oil Sands Access to Tidewater: A Political Risk Analysis of Pipeline Alternatives." *Canadian Public Policy / Analyse de Politiques* 39 (3): 371–91. https://dx.doi.org/10.3138/CPP.39.3.371

– 2021. *The Resistance Dilemma: Place-Based Movements and the Climate Crisis.* Cambridge, MA: MIT Press.

Hoberg, George, and Jeffrey Phillips. 2011. "Playing Defence: Early Responses to Conflict Expansion in the Oil Sands Policy Subsystem." *Canadian Journal of Political Science/Revue Canadienne de Science Politique* 44 (3): 507–27. https://doi.org/10.1017/S0008423911000473.

Hobson, Cole. 2013a. "SWN Displays Seismic Tests: Company Says There Are No Environmental or Safety Concerns from 2D Seismic Testing Beginning Soon." *Moncton Times and Transcript*, 31 May. Sec. Main.

– 2013b. "Anti-Shale Gas Court Case Could Work, Says Expert." *Moncton Times and Transcript*, 1 November. Sec. A.

– 2013c. "N.B. Has Safe Fracking History: Group." *Moncton Times and Transcript*, 9 November. Sec. A.

Hoppe, Robert. 2010. *The Governance of Problems: Puzzling, Powering and Participation.* Bristol, UK: Policy Press.

Houle, David, Erick Lachapelle, and Mark Purdon. 2015. "Comparative Politics of Sub-Federal Cap-and-Trade: Implementing the Western Climate Initiative." *Global Environmental Politics* 15 (3): 49–73. https://doi.org/10.1162/GLEP_a_00311.

Howarth, Robert W. 2014. "A Bridge to Nowhere: Methane Emissions and the Greenhouse Gas Footprint of Natural Gas." *Energy Science and Engineering*, May, n/a-n/a. https://doi.org/10.1002/ese3.35.

Howarth, Robert W., Renee Santoro, and Anthony Ingraffea. 2011. "Methane and the Greenhouse-Gas Footprint of Natural Gas from Shale Formations." *Climatic Change* 106 (4): 679–90. https://doi.org/10.1007/s10584-011-0061-5.

– 2012. "Venting and Leaking of Methane from Shale Gas Development: Response to Cathles et al." *Climatic Change* 113 (2): 537–49. https://doi.org/10.1007/s10584-012-0401-0.

Howe, Miles. 2015. *Debriefing Elsipogtog: The Anatomy of a Struggle.* Halifax: Fernwood Books.

Howell, Emily L., Nan Li, Heather Akin, Dietram A. Scheufele, Michael A. Xenos, and Dominique Brossard. 2017. "How Do U.S. State Residents Form Opinions about 'Fracking' in Social Contexts? A Multilevel Analysis." *Energy Policy* 106 (July): 345–55. https://doi.org/10.1016/j.enpol.2017.04.003.

Howlett, Michael. 2002. "Do Networks Matter? Linking Policy Network Structure to Policy Outcomes: Evidence from Four Canadian Policy Sectors 1990–2000." *Canadian Journal of Political Science / Revue Canadienne de Science Politique* 35 (2): 235–67. https://doi.org/10.1017/S0008423902778232

Howlett, Michael, M. Ramesh, and Anthony Perl. 2009. *Studying Public Policy: Policy Cycles and Policy Subsystems*. 3rd ed. Oxford: Oxford University Press.

Huber, John D., and Charles R. Shipan. 2002. *Deliberate Discretion? The Institutional Foundations of Bureaucratic Autonomy*. Cambridge: Cambridge University Press.

Hughes, J. David. 2015. "A Clear Look at BC LNG Energy Security, Environmental Implications and Economic Potential." Vancouver: CCPA. https://www.policyalternatives.ca/publications/reports/clear-look-bc-lng.

Huras, Adam. 2011. "Minister Considers Higher Shale Gas Royalty Rates; Bruce Northrup Said Current Percentage Is Too Low." *Moncton Times and Transcript*, 3 December. Sec. Main.

– 2012a. "N.B. Proposes Shale Gas Rules; Bruce Fitch Says Regulations Will Be 'the Strongest in North America.'" *Moncton Times and Transcript*, 18 May. Sec. Main.

– 2012b. "Shale Gas Company Already Meets Gov't Rules; But Corridor Resources Rep Says Some of Government's Rules Are Restrictive." *Moncton Times and Transcript*, 26 May.

– 2012c. "Havelock Audience Opposes Shale Gas; Consultation Tour on Shale Gas Development Continues." *Moncton Times and Transcript*, 19 June.

– 2012d. "Large Crowd Pans N.B. Shale Gas Development; Plans in Motion to Take Provincial Government to Court over Divisive Issue." *Moncton Times and Transcript*, 5 July. Sec. Main.

– 2012e. "Minister Applauds Energy Institute Recommendation; Proposed Institute Would Study Province's Developing Shale Gas Industry." *Moncton Times and Transcript*, 16 October. Sec. International.

– 2013a. "N.B. to Release Shale Gas Rules Today; Regulations Aimed at Setting Parameters for Development of Industry." *Moncton Times and Transcript*, 15 February. Sec. Main.

– 2013b. "New N.B. Shale Gas Regulations Are Questioned; Opposition Parties Worried about Compromise, Enforcement." *Moncton Times and Transcript*, 16 February. Sec. Job.

– 2013c. "Union Adds Voice to Call for Gas Moratorium." *Moncton Times and Transcript*, 15 November. Sec. C.

– 2013d. "First Nations Member of Energy Institute Resigns." *Moncton Times and Transcript*, 21 November.

– 2013e. "Embattled N.B. Energy Institute Regroups." *Moncton Times and Transcript*, 22 November. Sec. C.

Hussain, Yadullah. 2012. "Gas Industry Tackles Issues on 'Fracking'; Seeks to Lessen Fears with Better Transparency." *Calgary Herald*, 23 March. Sec. D.

ICF Consulting Canada. 2013. "The Future of Natural Gas Supply for Nova Scotia." Halifax: Nova Scotia Department of Energy. http://0-fs01.cito.gov .ns.ca.legcat.gov.ns.ca/deposit/b10664245.pdf.

Ingelson, Allan, and Tina Hunter. 2014. "Regulatory Comparison of Hydraulic Fracturing Fluid Disclosure Regimes in the United States, Canada, and Australia." *Natural Resources Journal* 54: 217.

Ingold, Karin, Manuel Fischer, and Paul Cairney. 2016. "Drivers for Policy Agreement in Nascent Subsystems: An Application of the Advocacy Coalition Framework to Fracking Policy in Switzerland and the UK." *Policy Studies Journal* 45 (3): 442–63. https://doi.org/10.1111/psj.12173.

Innovative Research Group. 2014. "Alberta Image and Market Access Survey Prepared for the Government of Alberta." Toronto. https://web.archive.org/web/20160326033707/http://www.alberta.ca/documents/Research-Alberta-Image-Market-Access-July2014.pdf.

Insights West. 2016. "British Columbians Oppose Fracking, Split on LNG Development." Vancouver: Insights West. https://web.archive.org/web/20160422030257/http://www.insightswest.com/news/british-columbians-oppose-fracking-split-on-lng-development/.

International Energy Agency (IEA). 2019. "World Energy Outlook 2019." Paris: International Energy Agency." https://www.iea.org/reports/world-energy-outlook-2019.

– 2021. "Net Zero by 2050: A Roadmap for the Global Energy Sector."

Interstate Oil and Gas Compact Commission (IOGCC). 2018. "Member States." Interstate Oil and Gas Compact Commission. https://web.archive.org/web/20080330001127/http://iogcc.publishpath.com/member-states.

– n.d. "IOGCC International: Achieving through Cooperation." Oklahoma City, OK: Interstate Oil and Gas Compact Commission. https://web.archive.org/web/20140513215151/http://iogcc.ok.gov/Websites/iogcc/Images/International%20English.pdf.

Jaccard, Marc, and Brad Griffin. 2010. "Shale Gas and Climate Targets: Can They Be Reconciled?" Vancouver: Pacific Institute for Climate Solutions. http://pics.uvic.ca/sites/default/files/uploads/publications/WP_Shale_Gas_and_Climate_Targets_August2010.pdf.

Jackson, Robert B., Avner Vengosh, J. William Carey, Richard J. Davies, Thomas H. Darrah, Francis O'Sullivan, and Gabrielle Pétron. 2014. "The Environmental Costs and Benefits of Fracking." *Annual Review of Environment and Resources* 39 (1): 327–62. https://doi.org/10.1146/annurev-environ-031113-144051.

Jacquet, Jeffrey B. 2012. "Landowner Attitudes toward Natural Gas and Wind Farm Development in Northern Pennsylvania." *Energy Policy* 50: 677–88. https://doi.org/10.1016/j.enpol.2012.08.011.

– 2014. "Review of Risks to Communities from Shale Energy Development." *Environmental Science and Technology* 48 (15): 8321–33. https://doi.org/10.1021/es404647x. Medline:24624971

Jacquet, Jeffrey B., and Richard C. Stedman. 2013. "Perceived Impacts from Wind Farm and Natural Gas Development in Northern Pennsylvania." *Rural Sociology* 78 (4): 450–72. https://doi.org/10.1111/ruso.12022

Janzwood, Amy. 2020. "Explaining Variation in Oil Sands Pipeline Projects." *Canadian Journal of Political Science/Revue Canadienne de Science Politique* 53 (3): 540–59. https://doi.org/10.1017/S0008423920000190.

– 2021. "The Contentious Politics of Mega Oil Sands Pipeline Projects." PhD diss., University of Toronto, Department of Political Science. https://tspace.library.utoronto.ca/bitstream/1807/105021/3/Janzwood _Amy_Alexandra_202103_PhD_thesis.pdf.

Janzwood, Amy, and Heather Millar. 2022. "Bridge Fuel Feuds: The Competing Interpretive Politics of Natural Gas in Canada." *Energy Research & Social Science* 88 (June): 102526. https://doi.org/10.1016/j.erss.2022 .102526.

Jasanoff, Sheila. 2003. "Technologies of Humility: Citizen Participation in Governing Science." *Minerva* 41 (3): 223–44.

Jaspal, Rusi, and Brigitte Nerlich. 2014. "Fracking in the UK Press: Threat Dynamics in an Unfolding Debate." *Public Understanding of Science* 23 (3): 348–63. https://doi.org/10.1177/0963662513498835. Medline:23942831

Jeakins, Paul. 2015. Personal Interview. 14 July. Victoria.

Jenkins-Smith, Hank C. 1988. "Analytical Debates and Policy Learning: Analysis and Change in the Federal Bureaucracy." *Policy Sciences* 21 (2–3): 169–211. https://doi.org/10.1007/BF00136407.

Jenkins-Smith, Hank C., Daniel Nohrstedt, Christopher M. Weible, and Paul A. Sabatier. 2014. "The Advocacy Coalition Framework: Foundations, Evolutions, and Ongoing Research." In *Theories of the Policy Process*, ed. Paul A. Sabatier and Christopher M. Weible. 3rd ed. . Boulder, CO: Westview Press.

Jenkins-Smith, Hank C., Carol L. Silva, Kuhika Gupta, and Joseph T. Ripberger. 2014. "Belief System Continuity and Change in Policy Advocacy Coalitions: Using Cultural Theory to Specify Belief Systems, Coalitions, and Sources of Change." *Policy Studies Journal* 42 (4): 484–508. https://doi .org/10.1111/psj.12071.

John, Peter. 2003. "Is There Life After Policy Streams, Advocacy Coalitions, and Punctuations: Using Evolutionary Theory to Explain Policy Change?" *Policy Studies Journal* 31 (4): 481–98. https://doi.org/10.1111/1541 -0072.00039.

Johnson, Elizabeth, and Laura Johnson. 2012. "Hydraulic Fracture Water Usage in Northeast British Columbia: Locations, Volumes and Trends." Geoscience Reports. Victoria: Ministry of Energy and Mines.

Jones, Bryan D. 1999. "Bounded Rationality." *Annual Review of Political Science* 2 (1): 297–321. https://doi.org/10.1146/annurev.polisci.2.1.297.

– 2017. "Behavioral Rationality as a Foundation for Public Policy Studies." *Cognitive Systems Research* 43 (June): 63–75. https://doi.org/10.1016/j.cogsys .2017.01.003.

Jordan, Andrew, and Elah Matt. 2014. "Designing Policies That Intentionally Stick: Policy Feedback in a Changing Climate." *Policy Sciences* 47 (3): 227–47. https://doi.org/10.1007/s11077-014-9201-x.

Kahan, Dan M. 2015. "The Politically Motivated Reasoning Paradigm, Part 1: What Politically Motivated Reasoning Is and How to Measure It." In *Emerging Trends in the Social and Behavioral Sciences*. New York: John Wiley and Sons. https://doi.org/10.1002/9781118900772.etrds0417.

Kahan, Dan M., Donald Braman, Geoffrey L. Cohen, John Gastil, and Paul Slovic. 2010. "Who Fears the HPV Vaccine, Who Doesn't, and Why? An Experimental Study of the Mechanisms of Cultural Cognition." *Law and Human Behavior* 34 (6): 501–16. https://doi.org/10.1007/s10979-009-9201-0 .Medline:20076997

Kahneman, Daniel. 2013. *Thinking, Fast and Slow*. Concord ON: Anchor Canada.

Kalaf-Hughes, Nicole, and Andrew R. Kear. 2018. "Framed for Compromise? The Role of Bill Framing in State Legislative Behavior on Natural Gas Policy." *Policy Studies Journal* 46: 598–628. https://doi-org.proxy.hil.unb.ca /10.1111/psj.12208.

Kanduth, Anna, and Jason Dion. 2022. "Electric Federalism: Policy for Aligning Canadian Electricity Systems with Net Zero." Ottawa, ON: Canadian Climate Institute. https://climateinstitute.ca/wp-content /uploads/2022/05/Electric-Federalism-May-4-2022.pdf.

Kasperson, Jeanne X., Roger E. Kasperson, Nick Pidgeon, and Paul Slovic. 2003. "The Social Amplification of Risk: Assessing Fifteen Years of Research and Theory." In *The Social Amplification of Risk*, ed. Nick Pidgeon, Roger E. Kasperson, and Paul Slovic, 13–46. Cambridge: Cambridge University Press.

Kasperson, Roger E., Ortwin Renn, Paul Slovic, Halina S. Brown, Jacque Emel, Robert Goble, Jeanne X. Kasperson, and Samuel Ratick. 1988. "The Social Amplification of Risk: A Conceptual Framework." *Risk Analysis* 8 (2): 177–87. https://doi.org/10.1111/j.1539-6924.1988.tb01168.x.

Kay, Adrian, and Phillip Baker. 2015. "What Can Causal Process Tracing Offer to Policy Studies? A Review of the Literature." *Policy Studies Journal* 43 (1): 1–21. https://doi.org/10.1111/psj.12092.

Keir, Jack. 2014. Personal interview conducted by phone December 9th 2014.

Kemfert, Claudia, Fabian Präger, Isabell Braunger, Franziska M. Hoffart, and Hanna Brauers. 2022. "The Expansion of Natural Gas Infrastructure Puts Energy Transitions at Risk." *Nature Energy* (July): 1–6. https://doi.org /10.1038/s41560-022-01060-3.

Kestler-D'Amours, Jillian. 2018. "Quebec to Ban Shale Gas Fracking, Tighten Rules for Oil and Gas Drilling." *CBC*, 6 June. https://www.cbc.ca/news /canada/montreal/quebec-fracking-ban-1.4694327.

Kingdon, John W. 1995. *Agendas, Alternatives, and Public Policies*. 2nd ed. New York: HarperCollins College Publishers.

Klinke, Andreas, and Ortwin Renn. 2012. "Adaptive and Integrative Governance on Risk and Uncertainty." *Journal of Risk Research* 15 (3): 273–92. https://doi.org/10.1080/13669877.2011.636838.

Kondash, Andrew, and Avner Vengosh. 2015. "Water Footprint of Hydraulic Fracturing." *Environmental Science and Technology Letters* 2 (10): 276–80. https://doi.org/10.1021/acs.estlett.5b00211.

Konschnik, Katherine, and Archana Dayalu. 2016. "Hydraulic Fracturing Chemicals Reporting: Analysis of Available Data and Recommendations for Policymakers." *Energy Policy* 88 (January): 504–14. https://doi.org/10.1016 /j.enpol.2015.11.002.

Koope, Bart, and Intrinsik. 2015. "Human Health Risk Assessment of Oil and Gas Activity in Northeastern British Columbia." Intrinsik. http://www2 .gov.bc.ca/assets/gov/health/keeping-bc-healthy-safe/human-health-risk -assessment-oil-and-gas/health-risk-assessment-oil-and-gas-ne-bc-phase -3-powerpoint-june2015.pdf.

Kriesky, J., B.D. Goldstein, K. Zell, and S. Beach. 2013. "Differing Opinions about Natural Gas Drilling in Two Adjacent Counties with Different Levels of Drilling Activity." *Energy Policy* 58 (Supplement C): 228–36. https://doi .org/10.1016/j.enpol.2013.03.005.

Lachapelle, Erick, and Simon Kiss. 2019. "Opposition to Carbon Pricing and Right-Wing Populism: Ontario's 2018 General Election." *Environmental Politics* 28 (5): 970–6. https://doi.org/10.1080/09644016.2019.1608659.

Lachapelle, Erick, Éric Montpetit, and Jean-Philippe Gauvin. 2014. "Public Perceptions of Expert Credibility on Policy Issues: The Role of Expert Framing and Political Worldviews." *Policy Studies Journal* 42 (4): 674–97. https://doi.org/10.1111/psj.12073.

Lamb, William F., Giulio Mattioli, Sebastian Levi, J. Timmons Roberts, Stuart Capstick, Felix Creutzig, Jan C. Minx, Finn Müller-Hansen, Trevor Culhane, and Julia K. Steinberger. 2020. "Discourses of Climate Delay." *Global Sustainability* 3: e17. https://doi.org/10.1017/sus.2020.13.

LaPierre, Louis. 2012. "The Path Forward." Fredericton: Government of New Brunswick. http://www2.gnb.ca/content/dam/gnb/Corporate/pdf /ShaleGas/en/ThePathForward.pdf.

Larkin, Patricia, Robert Gracie, Maurice Dusseault, and Daniel Krewski. 2018. "Ensuring Health and Environmental Protection in Hydraulic Fracturing: A Focus on British Columbia and Alberta, Canada." *The Extractive Industries and Society* 5 (4): 581–95. https://doi.org/10.1016/j.exis.2018.07.006.

Laroche, Jean. 2019. "McNeil Government's Fracking Ban Remains a Work in Progress." *CBC*, 23 April. https://www.cbc.ca/news/canada/nova-scotia /petroleum-resources-act-fracking-high-volume-hydraulic-fracturing -environment-1.5107870.

Lauer, Nancy E., Jennifer S. Harkness, and Avner Vengosh. 2016. "Brine Spills Associated with Unconventional Oil Development in North Dakota." *Environmental Science & Technology* 50 (10): 5389–97. https://doi.org /10.1021/acs.est.5b06349.

Lavoie, D., Z. Chen, N. Pinet, and S. Lyster. 2012. "A Review of November 24–25, 2011 Shale Gas Workshop, Calgary, Alberta – 1. Resource Evaluation Methodology." Open File 7088. Ottawa: Geological Survey of Canada. ftp://ftp2.cits.rncan.gc.ca/pub/geott/ess_pubs/290/290266 /of_7088.pdf.

Lee, Jeff. 2011. "B.C. to Continue 'Fracking' for Gas, despite Bans Elsewhere; Province's Energy and Mines Minister Confident with 'Maturity We Have in This Particular Field.'"*Vancouver Sun*, 30 March. Sec. BusinessBC.

Lee, Marc. 2015. "LNG and Employment in BC." Vancouver: CCPA. https://www.policyalternatives.ca/sites/default/files/uploads/publications /BC%20Office/2015/07/ccpa-bc_LNG_Employment_web.pdf.

– 2017. "BC's Natural Gas Giveaway: Production Soars, Revenues Plummet." CCPA. *Policy Note* (blog). 9 February. http://www.policynote.ca/giveaway/.

Leonard, Craig. 2014. Personal Interview. 18 November.

Lesch, Matthew. 2018. "Playing with Fiscal Fire: The Politics of Consumption Tax Reform." Phd diss., University of Toronto. https://tspace.library .utoronto.ca/handle/1807/87367.

– 2021. "Competing, Learning or Emulating? Policy Transfer and Sales Tax Reform." In *Provincial Policy Laboratories: Policy Diffusion and Transfer in Canada's Federal System*, ed. Brendan Boyd and Andrea Olive. Toronto: University of Toronto Press.

Lesch, Matthew, and Heather Millar. 2022. "Crisis, Uncertainty and Urgency: Processes of Learning and Emulation in Tax Policy Making." *West European Politics* 45 (4): 930–52. https://doi.org/10.1080/01402382.2021.1949681.

Levac, Leah R.E., and Sarah Marie Wiebe, eds. 2020. *Creating Spaces of Engagement: Policy Justice and the Practical Craft of Deliberative Democracy*. Toronto: University of Toronto Press.

Levin, Kelly, Benjamin Cashore, Steven Bernstein, and Graeme Auld. 2012. "Overcoming the Tragedy of Super Wicked Problems: Constraining Our Future Selves to Ameliorate Global Climate Change." *Policy Sciences* 45 (2): 123–52. https://doi.org/10.1007/s11077-012-9151-0.

Lewis, Eric. 2010. "Water Sale Raises Concerns; City of Moncton's Environment Committee to Examine Practice of Selling Municipal Water." *Moncton Times and Transcript*, 2 November. Sec. A. 1.

Lightfoot, Sheryl. 2020. "Unfinished Business: Implementation of the UN Declaration on the Rights of Indigenous Peoples in Canada." Montreal: Institute for Research on Public Policy. https://centre.irpp.org/research -studies/unfinished-business-implementation-of-the-un-declaration-on-the -rights-of-indigenous-peoples-in-canada/.

Lindblom, Charles. 1979. "Still Muddling, Not Yet Through." *Public Administration Review* 39 (6): 517–26.

– 1980. *Politics And Markets: The World's Political-Economic Systems*. New edition. New York: Basic Books.

Lindvall, Johannes. 2009. "The Real but Limited Influence of Expert Ideas." *World Politics* 61 (4): 703–30. https://doi.org/10.1017/S0043887109990104.

Lodge, Martin, and Kira Matus. 2014. "Science, Badgers, Politics: Advocacy Coalitions and Policy Change in Bovine Tuberculosis Policy in Britain." *Policy Studies Journal* 42 (3): 367–90. https://doi.org/10.1111/psj.12065.

Logan, Nick. 2014. "N.B. Election: Did Shale Gas and Fracking Sway the Vote?" *Global News*, 23 September.

Lowry, William R. 2008. "Disentangling Energy Policy from Environmental Policy*." *Social Science Quarterly* 89 (5): 1195–1211. https://doi.org/10.1111 /j.1540-6237.2008.00565.x.

Lowry, William R., and Mark Joslyn. 2014. "The Determinants of Salience of Energy Issues." *Review of Policy Research* 31 (3): 153–72. https://doi.org /10.1111/ropr.12069.

Lucas, Alastair R., Theresa Watson, and Kimmel Eric. 2014. "Regulating Multistage Hydraulic Fracturing: Challenges in a Mature Oil and Gas Jurisdiction." In *The Law of Energy Underground: Understanding New Developments in Subsurface Production, Transmission, and Storage*, ed. Donald N. Zillman, Aileen McHarg, Adrian Bradbrook, and Lila Barrera-Hernandez, 127–46. Oxford: Oxford University Press.

Macdonald, Douglas. 2020. *Carbon Province, Hydro Province: The Challenge of Canadian Energy and Climate Federalism*. Toronto: University of Toronto Press.

Macdonald, Douglas, and Matthew Lesch. 2015. "Management of Distributive Conflicts Impeding Expansion of Interprovincial Hydroelectricity Transmission." *Journal of Canadian Studies/Revue d'études Canadiennes* 49 (3): 191–221. https://doi.org/10.3138/jcs.49.3.191

MacDonald, Michael. 2011. "Everyone's Frothing about Fracking." *Halifax Chronicle Herald*, 11 December. http://thechronicleherald.ca /novascotia/41670-everyone%E2%80%99s -frothing-about-fracking.

– 2012. "Fracking Backwash Frustrates Oilman." *Halifax Chronicle Herald*, 3 October. http://thechronicleherald.ca/novascotia/142969-fracking -backwash-frustrates-oilman.

– 2014. "Two Studies in N.S. Say Fracking Poses Low Risk to Groundwater." *Halifax Chronicle Herald*, 3 June. http://thechronicleherald.ca/novascotia /1212034-two-studies-in-ns-say-fracking-poses-low-risk-to-groundwater.

MacIntyre, Mary Ellen. 2013. "Fracking Bylaw Wording under Pressure." *Halifax Chronicle Herald*, 21 January.

Magee, Shane. 2019. "Sussex-Area Fracking Plans Shelved over 'Regulatory Uncertainty.'" *CBC*, 13 August. https://www.cbc.ca/news/canada/new -brunswick/corridor-fracking-sussex-regulatory-uncertainty-1.5245024.

Maggetti, Martino, and Fabrizio Gilardi. 2016. "Problems (and Solutions) in the Measurement of Policy Diffusion Mechanisms." *Journal of Public Policy* 36 (1): 87–107. https://doi.org/10.1017/S0143814X1400035X.

Majone, Giandomenico. 1989. *Evidence, Argument, and Persuasion in the Policy Process*. New Haven: Yale University Press.

May, Peter J. 1992. "Policy Learning and Failure." *Journal of Public Policy* 12 (4): 331–54.

Mazerolle, Brent. 2011. "Search for Gas Stirs Debate; NBers Voicing Concerns over Controversial Hydraulic Fracturing Procedure." *Moncton Times and Transcript*, 27 May. Sec. A. 1.

– 2013a. "N.B. Has Huge Shale Gas Potential: LaPierre; Head of New Energy Institute Discusses Province's Future Energy Developments with Moncton Audience." *Moncton Times and Transcript*, 7 February. Sec. Main.

– 2013b. "Energy Institute Starts Work." *Moncton Times and Transcript*, 17 August. Sec. A.

Mazur, Allan. 2016. "How Did the Fracking Controversy Emerge in the Period 2010–2012?" *Public Understanding of Science* 25 (2): 207–22. https://doi.org /10.1177/0963662514545311.

McAdam, Doug, and Hilary Boudet. 2012. *Putting Social Movements in Their Place: Explaining Opposition to Energy Projects in the United States, 2000–2005*. Cambridge: Cambridge University Press.

McBeth, Mark K., Elizabeth A. Shanahan, Ruth J. Arnell, and Paul L. Hathaway. 2007. "The Intersection of Narrative Policy Analysis and Policy Change Theory." *Policy Studies Journal* 35 (1): 87–108. https://doi.org /10.1111/j.1541-0072.2007.00208.x.

McGowan, Francis. 2014. "Regulating Innovation: European Responses to Shale Gas Development." *Environmental Politics* 23 (1): 41–58. https://doi.org /10.1080/09644016.2012.740939.

McHardie, Daniel. 2016. "New Brunswick Extends Fracking Moratorium, Energy Minister Says." *CBC News*, 27 May. http://www.cbc.ca/news/canada /new-brunswick/arseneault-fracking-commission-report-1.3602849.

McKenzie, Janetta, and Angela V. Carter. 2021. "Stepping Stones to Keep Fossil Fuels in the Ground: Insights for a Global Wind down from Ireland." *The*

Extractive Industries and Society 8 (4): 101002. https://doi.org/10.1016/j
.exis.2021.101002.

McSheffrey, Elizabeth. 2015. "What You Need to Know about the Unist'ot'en-Pipeline Standoff." *Vancouver Observer*, 31 August. https://www
.vancouverobserver.com/news/what-you-need-know-about-unistoten-pipeline-standoff.

Mehta, Jal. 2010. "The Varied Roles of Ideas in Politics: From 'Whether' to 'How.'" In *Ideas and Politics in Social Science Research*, ed. Daniel Béland and Robert Henry Cox, 23–46. Oxford: Oxford University Press.

Meissner, Dirk. 2014. "Nisga'a Nation Signs LNG Pipeline Benefits Deal with B.C." *CBC News*, 21 November. http://www.cbc.ca/news/canada/british
-columbia/nisga-a-nation-signs-lng-pipeline-benefits-deal-with-b-c-1.2844672.

Melton, Noel, and Jotham Peters. 2013. "Provincial Carbon Tax Is Good for Your Wallet and Broader Economy." *Vancouver Sun*, 29 July. Sec. Letters.

Mildenberger, Matto. 2019. "New Brunswick's Timid Foray into Carbon Pricing." IRPP. https://policyoptions.irpp.org/fr/magazines/july-2019
/new-brunswicks-timid-foray-into-carbon-pricing/.

– 2020. *Carbon Captured: How Business and Labor Control Climate Politics*. American and Comparative Environmental Policy. Cambridge, MA: MIT Press.

Millar, Heather. 2013. "Comparing Accountability Relationships between Governments and Non-State Actors in Canadian and European International Development Policy." *Canadian Public Administration* 56 (2): 252–69. https://doi.org/10.1111/capa.12017.

– 2020. "Problem Uncertainty, Institutional Insularity, and Modes of Learning in Canadian Provincial Hydraulic Fracturing Regulation." *Review of Policy Research* 37 (6): 765–96. https://doi.org/10.1111/ropr.12401.

– 2021. "Interjurisdictional Transfer of Hydraulic Fracturing Regulations among Canadian Provinces." In *Provincial Policy Laboratories: Policy Transfer and Diffusion in Canada's Federal System.*, ed.Brendan Boyd and Andrea Olive, 49–68. Toronto: University of Toronto Press.

– Forthcoming. "Staying Put or Passing Through? The Politics of Deep Decarbonization in New Brunswick's Electricity Sector." In *The Higgs Years*, ed. Gabriel Arsenault. Montreal: McGill-Queen's University Press.

Millar, Heather, Eve Bourgeois, Steven Bernstein, and Matthew Hoffmann. 2021. "Self-Reinforcing and Self-Undermining Feedbacks in Subnational Climate Policy Implementation." *Environmental Politics* 30 (5): 791–810. https://doi.org/10.1080/09644016.2020.1825302.

Millar, Heather, Adrienne Davidson, and Linda A. White. 2020. "Puzzling Publics: The Role of Reflexive Learning in Universal Pre-Kindergarten (UPK) Policy Formulation in Canada and the US:" *Public Policy and Administration* 35 (3): 312–36. https://doi.org/10.1177
/0952076719889100.

Millar, Heather, Matthew Lesch, and Linda A. White. 2019. "Connecting Models of the Individual and Policy Change Processes: A Research Agenda." *Policy Sciences* 52 (1): 97–118. https://doi.org/10.1007/s11077 -018-9327-3.

Minkow, David. 2015. "Getting Off the Frack Track: How Anti-Fracking Campaigns Succeeded in New Brunswick and Nova Scotia." *Freshwater Alliance* (blog). 20 February. https://www.watercanada.net/feature/off-the -frack-track/.

Mirski, Pawel, and Len Coad. 2013. "Managing Expectations: Assessing the Potential of BC's Liquid Natural Gas Industry." Calgary: Canada West Foundation. https://web.archive.org/web/20131214172137 /http://cwf.ca/pdf-docs/publications/ManagingExpectations _October2013-2.pdf.

Moncton Times and Transcript. 2011a. "N.B. Gas Map Is Now Online." 10 November. Sec. Main.

– 2011b. "Protesters Oppose Hydrofracking; Hundreds Take to the Streets of Fredericton to Voice Concerns over Shale Gas Industry in N.B." 2 August. Sec. Main.

Montpetit, Éric. 2016. *In Defense of Pluralism: Policy Disagreement and Its Media Coverage.* Cambridge: Cambridge University Press.

Montpetit, Éric, and Erick Lachapelle. 2017. "Policy Learning, Motivated Scepticism, and the Politics of Shale Gas Development in British Columbia and Quebec." *Policy and Society* 36 (2): 195–214. https://doi.org/10.1080 /14494035.2017.1320846.

Montpetit, Éric, Erick Lachapelle, and Alexandre Harvey. 2016. "Advocacy Coalitions, the Media, and Hydraulic Fracturing in the Canadian Provinces of British Columbia and Quebec." In *Policy Debates on Hydraulic Fracturing: Comparing Coalition Politics in North America and Europe*, ed. Christopher M. Weible, Tanya Heikkila, Karin Ingold, and Manuel Fischer, 53–79. New York: Springer.

Moore, Dene. 2014. "Judge Pulls Plug on Green Groups' Fracking Suit." *Global News*, 16 October. http://globalnews.ca/news/1618526/judge-pulls-plug -on-green-groups-fracking-suit/.

Moore, Michele-Lee, Karena Shaw, Heather Castleden, Rosanna Breiddal, Megan Kot, and Mathew Murray. 2015. "Regional Snapshot Report Building Capacity to Build Trust: Key Challenges for Water Governance in Relation to Hydraulic Fracturing." Waterloo, ON: Canadian Water Network. https://cwn-rce.ca/wp-content/uploads/2014/04/Moore-et-al-2015- CWN-Report-Water-Governance-and-Hydraulic-Fracturing.pdf.

Morris, Chris. 2011. "Minister Investigating Shale Gas Exploration; Margaret Ann Blaney Headed to Pennsylvania next Week." *Moncton Times and Transcript*, 12 May. Sec. Main.

– 2014. "Nova Scotia Fracking Report Boosts Call for Moratorium in N.B." *Moncton Times and Transcript*, 29 July. Sec. Main.

Morrissy, John. 2009. "Sun Setting on Alberta's Natural Gas Empire." *Vancouver Sun*, 29 September. Sec. BusinessBC.

Morton, Brian, and Dan Healing. 2012. "New National, but Voluntary Rules for 'Fracking' Set by Industry Group; Best Operating Practices Align Well with Province's Own Regulation, Says Official with B.C. Oil and Gas Commission." *Vancouver Sun*, 13 January.

Moyson, Stéphane. 2018. "Policy Learning over a Decade or More and the Role of Interests Therein: The European Liberalization Policy Process of Belgian Network Industries." *Public Policy and Administration* 33 (1): 88–117. https://doi.org/10.1177/0952076716681206.

Moyson, Stéphane, Bastien Fievet, Maximilien Plancq, Sébastien Chailleux, and David Aubin. 2022. "Make it Loud and Simple: Coalition Politics and Problem Framing in the French Policy Process of Hydraulic Fracturing." *Review of Policy Research* 39 (4): 411–40. https://doi.org/10.1111/ropr.12473.

Moyson, Stéphane, Peter Scholten, and Christopher M. Weible. 2017. "Policy Learning and Policy Change: Theorizing Their Relations from Different Perspectives." *Policy and Society* 36 (2): 161–77. https://doi.org/10.1080/14494035.2017.1331879.

National Energy Board (NEB). 2009. "A Primer for Understanding Canadian Shale Gas." Calgary: Her Majesty the Queen in Right of Canada as Represented by the National Energy Board.

– 2011. "Tight Oil Developments in the Western Canada Sedimentary Basin." Ottawa: National Energy Board. https://web.archive.org/web/20140810200343/http://www.neb-one.gc.ca/clf-nsi/archives/rnrgynfmtn/nrgyrprt/l/tghtdvlpmntwcsb2011/tghtdvlpmntwcsb2011-eng.pdf.

National Energy Board (NEB), and Alberta Geological Survey. 2017. "Duvernay Resource Assessment Energy Briefing Note." Calgary: Her Majesty the Queen in Right of Canada as Represented by the National Energy Board. https://web.archive.org/web/20201031191141/https://www.cer-rec.gc.ca/en/data-analysis/energy-commodities/crude-oil-petroleum-products/report/2017-duvernay/2017dvrn-eng.pdf.

National Energy Board (NEB), and BC Ministry of Energy and Mines (BCMEM). 2011. "Ultimate Potential for Unconventional Natural Gas in Northeastern British Columbia's Horn River Basin." Calgary: National Energy Board.

Natural Resources Canada. 2015. "Shale and Tight Resources in Canada | Natural Resources Canada." http://www.nrcan.gc.ca/energy/sources/shale-tight-resources/17669.

– 2017. "Ontario's Shale and Tight Resources." https://www.nrcan.gc.ca/energy/sources/shale-tight-resources/17709.

Neville, Kate J. 2021. *Fueling Resistance: The Contentious Political Economy of Biofuels and Fracking*. Oxford: Oxford University Press.

Neville, Kate J., Jennifer Baka, Shanti Gamper-Rabindran, Karen Bakker, Stefan Andreasson, Avner Vengosh, Alvin Lin, Jewellord Nem Singh, and Erika Weinthal. 2017. "Debating Unconventional Energy: Social, Political, and Economic Implications." *Annual Review of Environment and Resources* 42 (1): 241–66. https://doi.org/10.1146/annurev-environ -102016-061102.

Neville, Kate J., and Erika Weinthal. 2016. "Mitigating Mistrust? Participation and Expertise in Hydraulic Fracturing Governance." *Review of Policy Research* 33 (6): 578–602. https://doi.org/10.1111/ropr.12201.

New Brunswick Commission on Hydraulic Fracturing. 2016. "New Brunswick Commission on Hydraulic Fracturing. Volume I: The Findings." Fredericton: New Brunswick Commission on Hydraulic Fracturing.

New Brunswick Department of Energy. 2011. "The New Brunswick Energy Blueprint." Fredericton: Department of Energy. https://www2.gnb.ca /content/dam/gnb/Departments/en/pdf/Publications/201110NBEnergy Blueprint.pdf.

New Brunswick Natural Gas Group. 2012a. "Responsible Environmental Management of Oil and Gas Activities in New Brunswick: List of References." Fredericton: Government of New Brunswick. https://www2.gnb.ca /content/dam/gnb/Corporate/pdf/ShaleGas/en/ListOfReferences.pdf.

– 2012b. "Responsible Environmental Management of Oil and Gas Activities in New Brunswick: Recommendations for Public Discussion." Fredericton: Government of New Brunswick. http://www2.gnb.ca/content/dam/gnb /Corporate/pdf/ShaleGas/en/RecommendationsDiscussion.pdf.

Nikiforuk, Andrew. 2015. *Slick Water: Fracking and One Insider's Stand against the World's Most Powerful Industry*. Vancouver: Greystone Books/David Suzuki Institute.

Noble, Tyler. 2012. "Nisga'a Declare Outright Opposition to Enbridge." *CFTKTV*, 19 January. http://www.cftktv.com/News /Story.aspx?ID=2184091.

Northrup, Bruce. 2014. Personal Interview. 6 November.

Nourallah, Laura. 2023. "Hydraulic Fracturing in New Brunswick: Trust, Deliberation, and Risk Decision-Making." In *Democratizing Risk Governance: Bridging Science, Expertise, Deliberation and Public Values*, ed. Monica Gattinger, 135–60. New York: Springer International Publishing. https://doi.org/10.1007/978-3-031-24271-7_6.

Nova Scotia Department of Energy. 2009. "Toward a Greener Future: Nova Scotia's 2009 Energy Strategy." Halifax, NS: Nova Scotia Department of Energy. https://web.archive.org/web/20161026215857/http://energy .novascotia.ca/sites/default/files/Energy-Strategy-2009.pdf.

– 2010. "Renewable Electricity Plan." Halifax: Government of Nova Scotia. http://energy.novascotia.ca/sites/default/files/renewable-electricity-plan.pdf.

– 2012. "Onshore Nova Scotia Petroleum Well Database." Halifax: Government of Nova Scotia. http://energy.novascotia.ca/sites/default /files/Onshore-NS-Petroleum-Well-Database.pdf.

Nova Scotia Fracking Resource and Action Coalition. n.d. "Member Organizations." *NOFRAC | | Nova Scotia Fracking Resource and Action Coalition* (blog). Accessed 17 November 2015. https://nofrac.wordpress.com /about/links-2/.

Nova Scotia Fracking Resource and Action Coalition, and Barb Harris. 2013. "Out of Control: Nova Scotia's Experience with Fracking for Shale Gas." Nova Scotia: NOFRAC. https://nofrac.wordpress.com/nofrac-reports /issue-paper-2/.

Nova Scotia New Democratic Party (NSNDP). 2009. "Better Deal 2009." Halifax: Nova Scotia New Democratic Party. https://www.poltext.org /sites/poltext.org/files/plateformesV2/Nouvelle-Ecosse/NS_PL_2009 _NDP_en.pdf.

Nowlin, Matthew C. 2011. "Theories of the Policy Process: State of the Research and Emerging Trends." *Policy Studies Journal* 39 (s1): 41–60. https://doi.org/10.1111/j.1541-0072.2010.00389_4.x.

O'Connor, Christopher D., and Kaitlin Fredericks. 2018. "Citizen Perceptions of Fracking: The Risks and Opportunities of Natural Gas Development in Canada." *Energy Research & Social Science* 42 (August): 61–9. https://doi .org/10.1016/j.erss.2018.03.005.

O'Connor, Joe. 2023. "Premier Frack: New Brunswick's Blaine Higgs Makes a Global Citizen's Case for Fracking." *Financial Post*, 29 May. https://financialpost.com/commodities/energy/oil-gas/new-brunswick -premier-blaine-higgs-natural-gas-fracking.

O'Donnell, Susan, Janice Harvey, Andrew Secord, Clive Baldwin, and J.P. Sapinski. 2023. "Contesting Energy Discourses through Action Research (CEDAR)." 25 June. https://cedar-project.org/.

Office of the Auditor General of Canada (OAGC). 2012. "Report of the Commissioner of the Environment and Sustainable Development: Chapter 5 Environmental Petitions." Ottawa: Office of the Auditor General of Canada.

Oil and Gas Activities Act, SBC. 2008. http://www.bclaws.ca/civix/document /id/complete/statreg/08036_01.

Olive, Andrea. 2016. "What Is the Fracking Story in Canada?" *The Canadian Geographer / Le Géographe Canadien*, January, 60: 32–45. https://doi.org /10.1111/cag.12257.

– 2018. "Oil Development in the Grasslands: Saskatchewan's Bakken Formation and Species at Risk Protection." *Cogent Environmental Science* 4 (1): 1443666. https://doi.org/10.1080/23311843.2018.1443666.

Olive, Andrea, and Ashlie B. Delshad. 2017. "Fracking and Framing: A Comparative Analysis of Media Coverage of Hydraulic Fracturing in

Canadian and US Newspapers." *Environmental Communication* 11 (6): 784–99. https://doi.org/10.1080/17524032.2016.1275734.

Olive, Andrea, and Katie Valentine. 2018. "Is Anyone out There? Exploring Saskatchewan's Civil Society Involvement in Hydraulic Fracturing." *Energy Research & Social Science* 39 (May): 192–7. https://doi.org/10.1016/j.erss.2017.11.014.

Palmer, Vaughn. 2008. "Victoria Strikes Gold in Sale of Gas Drilling Rights in Horn River Basin."*Vancouver Sun,* 6 March. Sec. News.

– 2011. "With LNG, B.C. Should 'Rush These Things.'" *Vancouver Sun,* 21 September. Sec. Westcoast News.

– 2013. "Dix Avoids Issues by Promising Studies; 'Wait and See' Tactic Used to Dodge Hot Topics." *Vancouver Sun,* 19 April. Sec. Westcoast News.

Papillon, Martin, and Thierry Rodon. 2017. "Indigenous Consent and Natural Resource Extraction Foundations for a Made-in-Canada Approach." IRPP Insight 16. Montreal: Institute for Research on Public Policy. http://irpp.org/wp-content/uploads/2017/07/insight-no16.pdf.

Paquet, Mireille, and Jörg Broschek. 2017. "This Is Not a Turn: Canadian Political Science and Social Mechanisms." *Canadian Journal of Political Science/Revue Canadienne de Science Politique* 50 (1): 295–310. https://doi.org/10.1017/S0008423917000038.

Parfitt, Ben. 2011. "Fracking Up Our Water, Hydro Power and Climate: BC's Reckless Pursuit of Shale Gas." Vancouver: Canadian Centre for Policy Alternatives.

– 2012. "Slow and Easy Will Win Energy Race; We Must Fully Understand the Interconnections and Interdependencies before Rushing to Any Decisions." *Vancouver Sun,* 14 February.

Parsons, Craig. 2016. "Ideas and Power: Four Intersections and How to Show Them." *Journal of European Public Policy* 23 (3): 446–63. https://doi.org/10.1080/13501763.2015.1115538.

Paterson, Matthew, Paul Tobin, and Stacy D. VanDeveer. 2022. "Climate Governance Antagonisms: Policy Stability and Repoliticization." *Global Environmental Politics* 22 (2): 1–11. https://doi.org/10.1162/glep_a_00647.

Pembina Institute. 2007. "Protecting Water, Producing Gas." Edmonton: Pembina Institute. https://www.pembina.org/reports/WaterGasFS.pdf.

– 2012. "Shale Gas in Canada. Towards Responsible Shale Gas Development in Canada: Opportunities and Challenges." Calgary: Pembina Institute. http://www.pembina.org/reports/sg-tlf-pre-read.pdf.

Penty, Rebecca. 2011a. "Living in Fear of Fracking; Rancher Concerned about Effects on Water Supply." *Calgary Herald,* 29 October. Sec. Calgary Business.

– 2011b. "Fracking Fears Spur Review of Oilpatch Regulations; Provinces Commited to Registry to Disclose Use of Chemicals." *Calgary Herald,* 30 December. Sec. Calgary Business.

Pidgeon, Nick, Merryn Thomas, Tristan Partridge, Darrick Evensen, and Barbara Herr Harthorn. 2017. "Hydraulic Fracturing: A Risk for Environment, Energy Security, and Affordability?" In *Risk Conundrums: Solving Unsolvable Problems*, ed. Roger E. Kasperson, 177–88. London: Routledge.

Pielke, Roger A. 2007. *The Honest Broker: Making Sense of Science in Policy and Politics.* Cambridge: Cambridge University Press.

Pierce, Jonathan J., Katrina Miller-Stevens, Isabel Hicks, Dova Castaneda Zilly, Saigopal Rangaraj, and Evan Rao. 2022. "How Anger and Fear Influence Policy Narratives: Advocacy and Regulation of Oil and Gas Drilling in Colorado." *Review of Policy Research* n/a: 1–23. https://doi.org/10.1111/ropr.12519.

Pierson, Paul. 1993. "When Effect Becomes Cause: Policy Feedback and Political Change." *World Politics* 45 (4): 595. https://doi.org/10.2307 /2950710.

– 2004. *Politics in Time: History, Institutions, and Social Analysis.* Princeton: Princeton University Press.

Pineau, Pierre-Olivier. 2021. "Improving Integration and Coordination of Provincially Managed Electricity Systems in Canada." Toronto: CICC, Canadian Institute for Climate Choices. https://climatechoices.ca /wp-content/uploads/2021/09/CICC-Improving-integration-and -coordination-of-provincially-managed-electricity-systems-in-Canada -by-Pierre-Olivier-Pineau-FINAL.pdf.

Poitras, Jacques. 2019a. "PCs Give Shale Gas Development Quiet Go-Ahead in Sussex Area." *CBC*, 4 June. https://www.cbc.ca/news/canada/new -brunswick/higgs-shale-gas-go-ahead-sussex-1.5162253.

– 2019b. "Higgs May Create His Own Carbon Tax in Wake of Federal Liberal Win." *CBC*, 22 October. https://www.cbc.ca/news/canada/new -brunswick/carbon-tax-new-brunswick-higgs-federal-election-1.5330765.

Pralle, Sarah. 2003. "Venue Shopping, Political Strategy, and Policy Change: The Internationalization of Canadian Forest Advocacy." *Journal of Public Policy* 23 (3): 233–60. https://doi.org/10.1017/S0143814X03003118.

– 2006a. *Branching Out, Digging In: Environmental Advocacy and Agenda Setting.* Washington, DC: Georgetown University Press.

– 2006b. "The 'Mouse That Roared': Agenda Setting in Canadian Pesticides Politics." *Policy Studies Journal* 34 (2): 171–94. https://doi.org/10.1111 /j.1541-0072.2006.00165.x.

– 2006c. "Timing and Sequence in Agenda-Setting and Policy Change: A Comparative Study of Lawn Care Pesticide Politics in Canada and the US." *Journal of European Public Policy* 13 (7): 987–1005. https://doi.org/10.1080 /13501760600923904.

Pralle, Sarah, and Jessica Boscarino. 2011. "Framing Trade-Offs: The Politics of Nuclear Power and Wind Energy in the Age of Global Climate Change." *Review of Policy Research* 28 (4): 323–46. https://doi.org/10.1111/j.1541 -1338.2011.00500.x.

Precht, Paul, and Don Dempster. 2012. "Jurisdictional Review of Hydraulic Fracturing Regulation: Report for Nova Scotia Hydraulic Fracturing Review Committee." Halifax: Nova Scotia Department of Energy and Nova Scotia Environment. http://www.novascotia.ca/nse /pollutionprevention/docs/Consultation.Hydraulic.Fracturing -Jurisdictional.Review.pdf.

Premier's Office. 2010. "Premier Lauds Historic Hydro Power Deal." Halifax: Government of Nova Scotia. http://novascotia.ca/news/release /?id=20101118001.

Prindle, David F. 2012. "Importing Concepts from Biology into Political Science: The Case of Punctuated Equilibrium." *Policy Studies Journal* 40 (1): 21–44. https://doi.org/10.1111/j.1541-0072.2011.00432.x.

Progressive Conservative Party of New Brunswick (PCNB). 2010. "Putting New Brunswick First...FOR A CHANGE." Fredericton: Progressive Conservative Party of New Brunswick. https://web.archive.org/web /20181119134713/https://www.poltext.org/sites/poltext.org/files /plateformes/nb2010pc_plt_en_13072011_132739.pdf.

Pryce, David. 2011. "N.B. Shale Gas: Addressing Challenges, Realizing Opportunities." *Moncton Times and Transcript*, May 3, 2011, sec. D. 6.

Pynn, Larry. 2013. "LNG Plants Will Scuttle Emissions Cuts; Liberals, NDP and Conservatives All Favour LNG Industry." *Vancouver Sun*, 9 May. Sec. News.

Rabe, Barry G. 2014. "Shale Play Politics: The Intergovernmental Odyssey of American Shale Governance." *Environmental Science and Technology*, February. https://doi.org/10.1021/es4051132.

– 2018. *Can We Price Carbon ?* Cambridge, MA: The MIT Press.

– 2022. "The Politics of Short-Lived Climate Pollutants and North American Methane Policy." Ann Arbor, MI: University of Michigan, Ford School of Public Policy. https://fordschool.umich.edu/sites/default/files/2022-04 /NACP_Rabe_final.pdf.

Rabe, Barry G., and Christopher Borick. 2013. "Conventional Politics for Unconventional Drilling? Lessons from Pennsylvania's Early Move into Fracking Policy Development." *Review of Policy Research* 30 (3): 321–40. https://doi.org/10.1111/ropr.12018.

Radaelli, Claudio M. 2009. "Measuring Policy Learning: Regulatory Impact Assessment in Europe." *Journal of European Public Policy* 16 (8): 1145–64. https://doi.org/10.1080/13501760903332647.

Rahm, Dianne. 2011. "Regulating Hydraulic Fracturing in Shale Gas Plays: The Case of Texas." *Energy Policy* 39 (5): 2974–81. https://doi.org/10.1016/j .enpol.2011.03.009.

Raimi, Daniel. 2018. *The Fracking Debate: The Risks, Benefits, and Uncertainties of the Shale Revolution.* New York: Columbia University Press. https://doi.org /10.7312/raim18486.

Rayher, Fiona, and Damien Gillis, dirs. 2015. *Fractured Land*. Documentary. Two Island Films, Moving Images Distribution. http://www.fracturedland.com.

Raymond, Leigh. 2019. "Ontario's Carbon Price Experience Is a Cautionary Tale." IRPP. *Policy Options* (blog). 10 July. https://policyoptions.irpp.org /magazines/july-2019/ontarios-carbon-price-experience-is-a-cautionary-tale/.

Renn, Ortwin. 2008. *Risk Governance: Coping with Uncertainty in a Complex World*. London: Earthscan.

Repetto, Robert C., ed. 2006. *Punctuated Equilibrium and the Dynamics of U.S. Environmental Policy*. New Haven: Yale University Press.

Richardson, Nathan, Madeline Gottlieb, Alan Krupnick, and Hannah Wiseman. 2013. "The State of State Shale Gas Regulation." Washington, DC: Resources For the Future. https://web.archive.org/web/20130820135553 /http://www.rff.org/rff/documents/RFF-Rpt-StateofStateRegs_Report.pdf.

Rietig, Katharina. 2018. "The Links among Contested Knowledge, Beliefs, and Learning in European Climate Governance: From Consensus to Conflict in Reforming Biofuels Policy." *Policy Studies Journal* 46 (1): 137–59. https://doi .org/10.1111/psj.12169.

Rinfret, Sara, Jeffrey J. Cook, and Michelle C. Pautz. 2014. "Understanding State Rulemaking Processes: Developing Fracking Rules in Colorado, New York, and Ohio." *Review of Policy Research* 31 (2): 88–104. https://doi .org/10.1111/ropr.12060.

Rivard, Christine, Denis Lavoie, René Lefebvre, Stephan Séjourné, Charles Lamontagne, and Mathieu Duchesne. 2014. "An Overview of Canadian Shale Gas Production and Environmental Concerns." In "Environmental Geology and the Unconventional Gas Revolution." Special Issue. *International Journal of Coal Geology* 126 (June): 64–76. https://doi.org /10.1016/j.coal.2013.12.004.

Rivard, Christine, J.W. Molson, D.J. Soeder, E.G. Johnson, S.E. Grasby, B. Wang, and A. Rivera. 2012. "A Review of the November 24–25, 2011 Shale Gas Workshop, Calgary Alberta – 2. Groundwater Resources." Open File 7096. Ottawa: Geological Survey of Canada. https://publications.gc.ca /collections/collection_2016/rncan-nrcan/M183-2-7096-eng.pdf.

Rokosh, C.D., S. Lyster, S.D.A. Anderson, A.P. Beaton, H. Berhane, T Brazzoni, D. Chen, et al. 2012. "Summary of Alberta's Shale- and Siltstone-Hosted Hydrocarbon Resource Potential." Edmonto: Energy Resources Conservation Board.

Rokosh, C.D., J.G. Pawlowicz, H. Berhane, S.D.A. Anderson, and A.P. Beaton. 2009. "What Is Shale Gas? An Introduction to Shale-Gas Geology in Alberta." ERCB/AGS Open File Report 2008–08. Edmonton: Energy Resources Conservation Board.

Rose, Richard. 1991. "What Is Lesson-Drawing?" *Journal of Public Policy* 11 (1): 3–30. https://doi.org/10.1017/S0143814X00004918.

Ross, Selena. 2013. "The Election and the Environment." *Halifax Chronicle Herald*, 4 October. http://thechronicleherald.ca/novascotia/1158527-the -election-and-the-environment.

– 2014a. "Fracking Review Members Selected." *Halifax Chronicle Herald*, 4 February. http://thechronicleherald.ca/novascotia/1184435-fracking -review-members-selected.

– 2014b. "N.S. Fracking Review Will Be Exhaustive – Panel." *Halifax Chronicle Herald*, 13 February. http://thechronicleherald.ca/novascotia/1186807-ns -fracking-review-will-be-exhaustive-panel.

– 2014c. "Fractious Debate a Theme at Halifax Fracking Meeting." *Halifax Chronicle Herald*, 15 April. http://thechronicleherald.ca/metro/1200767 -fractious-debate-a-theme-at-halifax-fracking-meeting.

Sabatier, Paul A. 1988. "An Advocacy Coalition Framework of Policy Change and the Role of Policy-Oriented Learning Therein." *Policy Sciences* 21 (2–3): 129–68. https://doi.org/10.1007/BF00136406.

Sabatier, Paul A, and Christopher M. Weible. 2007. "The Advocacy Coalition Framework: Innovations and Clarifications." In *Theories of the Policy Process*, ed. Paul A Sabatier, 189–222. 2nd ed. Boulder: Westview Press.

Salas, C.J., and D. Murray. 2013. "Developing a Water Monitoring Network in the Horn River Basin, Northeastern British Columbia (Parts of NTS 0941, J, O, P)." Victoria: Geoscience BC. http://www.geosciencebc .com/i/pdf/SummaryofActivities2012/SoA2012_Salas_Water _Monitoring.pdf.

Salomons, Geoffrey H., and George Hoberg. 2014. "Setting Boundaries of Participation in Environmental Impact Assessment." *Environmental Impact Assessment Review* 45 (February): 69–75. https://doi.org/10.1016/j .eiar.2013.11.001.

Scharpf, Fritz W. 1989. "Decision Rules, Decision Styles and Policy Choices." *Journal of Theoretical Politics* 1 (2): 149–76. https://doi.org/10.1177 /0951692889001002003.

Schattschneider, Elmer Eric. 1960. *The Semisovereign People: A Realist's View of Democracy in America*. New York: Holt, Rinehart and Winston.

Schenk, Olga, Michelle H.W. Lee, Naveed H. Paydar, John A. Rupp, and John D. Graham. 2014. "Unconventional Gas Development in the U.S. States: Exploring the Variation." European Journal of Risk Regulation. December. https://doi.org/10.1017/S1867299X00004050.

Schmidt, Vivien A. 2008. "Discursive Institutionalism: The Explanatory Power of Ideas and Discourse." *Annual Review of Political Science* 11: 303–26. https://doi.org/10.1146/annurev.polisci.11.060606.135342

– 2013. "Democracy and Legitimacy in the European Union Revisited: Input, Output and 'Throughput.'" *Political Studies* 61 (1): 2–22. https://doi .org/10.1111/j.1467-9248.2012.00962.x.

Schmidt, Vivien A. and Matthew Wood. 2019. "Conceptualizing Throughput Legitimacy: Procedural Mechanisms of Accountability, Transparency, Inclusiveness and Openness in EU Governance." *Public Administration* 97 (4): 727–40. https://doi.org/10.1111/padm.12615.

Schon, Donald A., and Martin Rein. 1995. *Frame Reflection: Toward the Resolution of Intractable Policy Controversies*. New York: Basic Books.

Schultz, R., G. Atkinson, D. W. Eaton, Y. J. Gu, and H. Kao. 2018. "Hydraulic Fracturing Volume Is Associated with Induced Earthquake Productivity in the Duvernay Play." *Science* 359 (6373): 304–8. https://doi.org/10.1126/science.aao0159. Medline:29348233

Seto, Karen C., Steven J. Davis, Ronald B. Mitchell, Eleanor C. Stokes, Gregory Unruh, and Diana Ürge-Vorsatz. 2016. "Carbon Lock-In: Types, Causes, and Policy Implications." *Annual Review of Environment and Resources* 41 (1): 425–52. https://doi.org/10.1146/annurev-environ-110615-085934.

Shaffer, Blake. 2021. "Technical Pathways to Aligning Canadian Electricity Systems with Net Zero Goals." Toronto: CICC, Canadian Institute for Climate Choices. https://climatechoices.ca/wp-content/uploads/2021/09/CICC-Technical-pathways-to-aligning-Canadian-electricity-systems-with-net-zero-goals-by-Blake-Shaffer-FINAL-1.pdf.

Sierra Club Atlantic Chapter. 2011. "Petition to Stop Fracking in Nova Scotia | Sierra Club Canada." 22 March. http://www.sierraclub.ca/en/node/2534.

Sierra Club BC. 2018. "Fossil Fuels." *Sierra Club BC* (blog). https://web.archive.org/web/20181228145629/https://sierraclub.bc.ca/campaigns/fossil-fuels/.

Simmons, Beth A., Frank Dobbin, and Geoffrey Garrett, eds. 2008. *The Global Diffusion of Markets and Democracy*. 1st ed. Cambridge: Cambridge University Press.

Simon, Herbert A. 1985. "Human Nature in Politics: The Dialogue of Psychology with Political Science." *American Political Science Review* 79 (2): 293–304. https://doi.org/10.2307/1956650

Simpson, Jeffrey, Mark Jaccard, and Nic Rivers. 2008. *Hot Air: Meeting Canada's Climate Change Challenge*. Revised, Updated ed. Toronto: Emblem Editions.

Simpson, Scott. 2006. "Gas Rights to Net $2 Billion: New Technology Allows Year-Round Drilling in Northeast: [Final Edition]." *Vancouver Sun*, 23 June. Sec. BusinessBC.

– 2008. "B.C. Has Potential to Be a Leader in Natural Gas; Demand Is on the Rise, and the Province Is Getting Set to Cash In." *Vancouver Sun*, 13 December. Sec. Businessbc.

– 2009a. "B.C. Collects $2.4 Billion in Leases from Gas and Oil Companies; 'Leading Economic Engine in the Province,' Energy Minister Says."*Vancouver Sun*, 27 March. Sec. Businessbc.

– 2009b. "Incentives Aim to Boost Gas Exploration in Northeastern B.C.; Stimulus Package Designed to Encourage Growth despite Relatively Low Gas Prices." *Vancouver Sun*, 7 August. Sec. C.

– 2010. "B.C. Energy Resource Has Its Ups and Downs in 2010; Gas Prices Down but Economic Activity on Upswing in Northeast B.C." *Vancouver Sun*, 26 December. Sec. BusinessBC.

Skogstad, Grace. 2003. "Who Governs? Who Should Govern? Political Authority and Legitimacy in Canada in the Twenty-First Century." *Canadian Journal of Political Science/Revue Canadienne de Science Politique* 36 (5): 955–73. https://doi.org/10.1017/S0008423903778925.

– 2008. "Policy Networks and Policy Communities: Conceptualizing State-Societal Relationships in the Policy Process." In *The Comparative Turn in Canadian Political Science*, ed. Linda A. White, Richard Simeon, Robert Vipond, and Jennifer Wallner, 205–20. Vancouver: UBC Press.

– ed. 2011a. *Policy Paradigms, Transnationalism, and Domestic Politics*. Toronto: University of Toronto Press.

– 2011b. "Contested Accountability Claims and GMO Regulation in the European Union." *JCMS: Journal of Common Market Studies* 49 (4): 895–915. https://doi.org/10.1111/j.1468-5965.2010.02166.x.

Slovic, Paul. 1987. "Perception of Risk." *Science* 236 (4799): 280–5.

– 1993. "Perceived Risk, Trust, and Democracy." *Risk Analysis* 13 (6): 675–82. https://doi.org/10.1111/j.1539-6924.1993.tb01329.x.

– 2000. *The Perception of Risk*. London: Earthscan.

Small, Mitchell J., Paul C. Stern, Elizabeth Bomberg, Susan M. Christopherson, Bernard D. Goldstein, Andrei L. Israel, Robert B. Jackson, et al. 2014. "Risks and Risk Governance in Unconventional Shale Gas Development." *Environmental Science and Technology* 48 (15): 8289–97. https://doi.org/10.1021/es502111u. Medline:24983403

Soroka, Stuart, and Christopher Wlezien. 2010. *Degrees of Democracy: Politics, Public Opinion, and Policy*. New York: Cambridge University Press.

Souther, Sara, Morgan W. Tingley, Viorel D. Popescu, David TS Hayman, Maureen E. Ryan, Tabitha A. Graves, Brett Hartl, and Kimberly Terrell. 2014. "Biotic Impacts of Energy Development from Shale: Research Priorities and Knowledge Gaps." *Frontiers in Ecology and the Environment* 12 (6): 330–8. https://doi.org/10.1890/130324.

Sovacool, Benjamin K. 2014. "Cornucopia or Curse? Reviewing the Costs and Benefits of Shale Gas Hydraulic Fracturing (Fracking)." *Renewable and Sustainable Energy Reviews* 37 (September): 249–64. https://doi.org/10.1016/j.rser.2014.04.068.

Sovacool, Benjamin K., Patrick Schmid, Andy Stirling, Goetz Walter, and Gordon MacKerron. 2020. "Differences in Carbon Emissions Reduction between Countries Pursuing Renewable Electricity versus Nuclear Power." *Nature Energy* 5 (11): 928–35. https://doi.org/10.1038/s41560-020-00696-3.

Stanec Consulting Ltd. 2014. "Phased Environmental Impact Assessment (EIA) Submission for Development of the Bronson Hydrocarbon Well: Well Pad Construction and Vertical Stratigraphic Drilling."

Statistics Canada. 2011. "Table A.34 Gross Domestic Product per Capita, Canada, Provinces and Territories, 2005/2006 to 2009/2010 (in Current Dollars)." Ottawa: Statistics Canada. http://www.statcan.gc.ca/pub/81 -595-m/2011095/tbl/tbla.34-eng.htm.

– 2014. "Electric Power Generation, by Class of Electricity Producer, Annual (Megawatt Hour), Table 127-0007." Ottawa: Statistics Canada. https://web .archive.org/web/20170614142214/http://www5.statcan.gc.ca/cansim /a26?lang=eng&id=1270007.

– 2018. "Population Estimates, Quarterly." 27 June. Ottawa: Statistics Canada. https://www150.statcan.gc.ca/t1/tbl1/en/tv.action?pid =1710000901.

Stephan, Hannes R. 2017. "The Discursive Politics of Unconventional Gas in Scotland: Drifting towards Precaution?" *Energy Research and Social Science* 23 (January): 159–68. https://doi.org/10.1016/j.erss.2016.09.006.

– 2020. "Shaping the Scope of Conflict in Scotland's Fracking Debate: Conflict Management and the Narrative Policy Framework." *Review of Policy Research* 37 (1): 64–91. https://doi.org/10.1111/ropr.12365.

Stephenson, Eleanor, Alexander Doukas, and Karena Shaw. 2012. "'Greenwashing Gas: Might a "Transition Fuel" Label Legitimize Carbon-Intensive Natural Gas Development?'" *Energy Policy* 46 (July): 452–9. https://doi.org/10.1016/j.enpol.2012.04.010.

Stephenson, Eleanor, and Karena Shaw. 2013. "A Dilemma of Abundance: Governance Challenges of Reconciling Shale Gas Development and Climate Change Mitigation." *Sustainability* 5 (5): 2210–32. https://doi.org/10.3390 /su5052210.

Stokes, Leah C. 2013. "The Politics of Renewable Energy Policies: The Case of Feed-in Tariffs in Ontario, Canada." *Energy Policy* 56 (May): 490–500. https://doi.org/10.1016/j.enpol.2013.01.009

– 2016. "Electoral Backlash against Climate Policy: A Natural Experiment on Retrospective Voting and Local Resistance to Public Policy." *American Journal of Political Science* 60 (4): 958–74. https://doi.org/10.1111/ajps.12220.

Stone, Deborah. 1989. "Causal Stories and the Formation of Policy Agendas." *Political Science Quarterly* 104 (2): 281–300. https://doi.org/10.2307/2151585.

– 2011. *The Policy Paradox*. New York: W.W. Norton.

Stone, Diane. 2017. "Understanding the Transfer of Policy Failure: Bricolage, Experimentalism and Translation." *Policy and Politics* 45 (1): 55–70. https://doi.org/10.1332/030557316X14748914098041.

Summers, Ken. 2016. "Fracking Company Leaves Province with the Cleanup." *Nova Scotia Advocate* (blog). 14 December. https://nsadvocate .org/2016/12/14/fracking-company-leaves-province-with-the-cleanup/.

Sunstein, Cass R. 2005. *Laws of Fear: Beyond the Precautionary Principle*. 1st ed. Cambridge: Cambridge University Press.

– 2009. *Worst-Case Scenarios*. Cambridge, MA: Harvard University Press.

Sutton, Jeannette, and Shari R. Veil. 2017. "Risk Communication and Social Media." In *Risk Conundrums: Solving Unsolvable Problems*, ed. Roger E. Kasperson, 96–111. London: Routledge.

Taber, Charles. 2003. "Information Processing and Public Opinion." In *Oxford Handbook of Political Psychology*, ed. David O. Sears, Leonie Huddy, and Robert Jervis, 433–76. Oxford: Oxford University Press.

Taylor, Roger. 2012. "Fracking Controversy Won't Just Go Away on Its Own." *Halifax Chronicle Herald*, 17 April. http://thechronicleherald.ca /business/87709-fracking-controversy-wont-just-go-away-on-its-own.

– 2013. "Deep Panuke Still Afloat." *Halifax Chronicle Herald*, 5 November. http://thechronicleherald.ca/business/1165379-taylor-deep-panuke-still-afloat.

Taylor-Gooby, Peter, and Jens Zinn, eds. 2006. *Risk in Social Science*. Oxford: Oxford University Press.

Terrace Standard. 2012. "Nisga'a Add to Enbridge Opposition." *Terrace Standard*, 19 January. https://www.terracestandard.com/news/nisgaa -add-to-enbridge-opposition-6015311.

Thorn, Adam. 2018. "Issue Definition and Conflict Expansion: The Role of Risk to Human Health as an Issue Definition Strategy in an Environmental Conflict." *Policy Sciences* 51 (1): 59–76. https://doi.org/10.1007/s11077-018 -9312-x.

Toogood, Allison. 2013. "Elsipogtog Chief, Council Oppose Shale Gas Testing; Police Called in as Protesters Confront SWN Vehicle." *Moncton Times and Transcript*, 5 June. Sec. Main.

Trein, Philipp. 2015. "Literature Report: A Review of Policy Learning in Five Strands of Political Science Research." D.51. Working Paper Series 2015. Brussels: INSPIRES Project: European Commission. http://www.inspires -research.eu/userfiles/D5_1%20Review%20Essay(1).pdf.

– 2018. "Median Problem Pressure and Policy Learning: An Exploratory Analysis of European Countries." In *Learning in Public Policy: Analysis, Modes and Outcomes*, ed. Claire A. Dunlop, Claudio M. Radaelli, and Philipp Trein, 243–66. International Series on Public Policy. New York: Springer International Publishing. https://doi.org/10.1007/978-3-319 -76210-4_11.

Triangle Petroleum Corporation. 2015. "About | Triangle Petroleum Corporation." http://www.trianglepetroleum.com/about.

Turnbull, Lori Beth. 2009. "The Nova Scotia Provincial Election of 2009." *Canadian Political Science Review* 3 (3): 69–76. https://doi.org/10.24124 /c677/2009151.

Tversky, A., and D. Kahneman. 1981. "The Framing of Decisions and the Psychology of Choice." *Science* 211 (4481): 453–8. https://doi.org/10.1126 /science.7455683. Medline:7455683

Urquhart, Ian. 2018. *Costly Fix: Power, Politics, and Nature in the Tar Sands*. Toronto: University of Toronto Press.

US Energy Information Administration (US EIA). 2013a. "Annual Energy Outlook 2013 Early Release Overview." Washington, DC: US Department of Energy. http://www.eia.gov/forecasts/aeo/er/index.cfm.

– 2013b. "Technically Recoverable Shale Oil and Shale Gas Resources: An Assessment of 137 Shale Formations in 41 Countries Outside the United States." Washington, DC: US Department of Energy. http://www.eia.gov /analysis/studies/worldshalegas/.

– 2014. "Annual Energy Outlook 2014 Early Release Overview." Washington, DC: US Department of Energy. http://www.eia.gov/forecasts/aeo/.

– 2015. "Size of Assessed Shale Gas and Shale Oil Resources, at Basin- and Formation-Levels." Washington, DC: US Department of Energy. https://www.eia.gov/analysis/studies/worldshalegas/.

– 2017. "Annual Energy Outlook 2017." Washington, DC: US Department of Energy. https://www.eia.gov/outlooks/aeo/.

– 2018a. "Henry Hub Natural Gas Spot Price (Dollars per Million Btu)." Washington, DC: US Department of Energy. http://www.eia.gov/dnav /ng/hist/rngwhhdd.htm.

– 2018b. "What Are Ccf, Mcf, Btu, and Therms? How Do I Convert Natural Gas Prices in Dollars per Ccf or Mcf to Dollars per Btu or Therm?" Washington, DC: US Department of Energy. https://www.eia.gov/tools /faqs/faq.php?id=45&t=8.

– 2019. "EIA – Annual Energy Outlook 2019." Washington, DC: US Department of Energy. https://www.eia.gov/outlooks/aeo/.

– 2020. "The Distribution of U.S. Oil and Natural Gas Wells by Production Rate." Washington, DC: US Department of Energy.https://www.eia.gov /petroleum/wells/pdf/full_report.pdf.

– 2022. "Annual Energy Outlook 2022." Washington, DC: US Department of Energy. https://www.eia.gov/outlooks/aeo/narrative/production/sub -topic-01.php.

Vancouver Sun. 2011. "LNG Plants Essential to Develop B.C.'s Natural Gas Bounty." 23 September. Sec. Editorial.

– 2012. "B.C.'s LNG Exports a Good News Story." 26 May.

– 2013. "The Natural Gas Dialogue." 30 May. Sec. BusinessBC.

VanNijnatten, Debora, and Douglas Macdonald. 2020. "Canada and the Climate Policy Dilemma." In *Canadian Politics*, ed. James Bickerton and Alain-G. Gagnon. 7th ed. Toronto: University of Toronto Press.

Vasi, Ion Bogdan, Edward T. Walker, John S. Johnson, and Hui Fen Tan. 2015. "'No Fracking Way!' Documentary Film, Discursive Opportunity, and Local Opposition against Hydraulic Fracturing in the United States, 2010 to 2013." *American Sociological Review* 80 (5): 934–59. https://doi.org /10.1177/0003122415598534.

Vengosh, Avner, Robert B. Jackson, Nathaniel Warner, Thomas H. Darrah, and Andrew Kondash. 2014. "A Critical Review of the Risks to Water Resources from Unconventional Shale Gas Development and Hydraulic Fracturing in the United States." *Environmental Science and Technology* 48 (15): 8334–48. https://doi.org/10.1021/es405118y. Medline:24606408

Vogel, David. 2012. *The Politics of Precaution: Regulating Health, Safety, and Environmental Risks in Europe and the United States.* Princeton, NJ: Princeton University Press.

Volpé, Jeannot, and William M. Thompson. 2011. "Final Report; New Brunswick Energy Commission 2010–2011." New Brunswick: New Brunswick Energy Commission. https://web.archive.org/web/20170628170603/http://www .atlanticaenergy.org/pdfs/natural_gas/Safety/NBEnergy CommissionFinalReport_2011.pdf.

Waldron, Ingrid R.G. 2018. *There's Something in the Water: Environmental Racism in Indigenous & Black Communities.* Black Point, NS: Fernwood Publishing.

Walker, Chad, and Jamie Baxter. 2017. "Procedural Justice in Canadian Wind Energy Development: A Comparison of Community-Based and Technocratic Siting Processes." *Energy Research and Social Science* 29 (July): 160–9. https://doi.org/10.1016/j.erss.2017.05.016.

Walker, Chad, Stacia Ryder, Jean-Pierre Roux, Zoé Chateau, and Patrick Devine-Wright. 2023. "Chapter 34 – Contested Scales of Democratic Decision-Making and Procedural Justice in Energy Transitions." In *Energy Democracies for Sustainable Futures*, ed. Majia Nadesan, Martin J. Pasqualetti, and Jennifer Keahey, 317–26. Cambridge MA: Academic Press. https://doi.org/10.1016/B978-0-12-822796-1.00034-6.

Walsh, Brian. 2014. "The Seismic Link between Fracking and Earthquakes." *Time Magazine*, 1 May. http://time.com/84225/fracking-and-earthquake -link/.

Warner, Barbara, and Jennifer Shapiro. 2013. "Fractured, Fragmented Federalism: A Study in Fracking Regulatory Policy." *Publius: The Journal of Federalism* 43 (3): 474–96. https://doi.org/10.1093/publius/pjt014.

Weible, Christopher M. 2008. "Expert-Based Information and Policy Subsystems: A Review and Synthesis." *Policy Studies Journal* 36 (4): 615–35. https://doi.org/10.1111/j.1541-0072.2008.00287.x.

Weible, Christopher M., and Tanya Heikkila. 2016. "Comparing the Politics of Hydraulic Fracturing in New York, Colorado, and Texas." *Review of Policy Research* 33 (3): 232–50. https://doi.org/10.1111/ropr.12170.

Weible, Christopher M., and Daniel Nohrstedt. 2012. "The Advocacy Coalition Framework: Coalitions, Learning and Policy Change." In *Routledge Handbook of Public Policy*, ed. Eduardo Araral, Scott Fritzen, Michael Howlett, M. Ramesh, and Xun Wu, 125–37. London: Routledge.

Wellstead, Adam, Paul Cairney, and Kathryn Oliver. 2018. "Reducing Ambiguity to Close the Science-Policy Gap." *Policy Design and Practice* 1 (2): 115–25. https://doi.org/10.1080/25741292.2018.1458397.

Weston, Greg. 2010. "Greens to Halt Fracking If Elected: Specific Rules Needed to Address Risks of Natural Gas Exploration: Party." *Moncton Times and Transcript*, 24 August.

Weyland, Kurt. 2005. "Theories of Policy Diffusion: Lessons from Latin American Pension Reform." *World Politics* 57 (2): 262–95. https://doi.org /10.1353/wp.2005.0019

Wheeler, David, Margo MacGregor, Frank Atherton, Kevin Christmas, Shawn Dalton, Maurice Dusseault, Graham Gagnon, et al. 2015. "Hydraulic Fracturing – Integrating Public Participation with an Independent Review of the Risks and Benefits." *Energy Policy* 85 (October): 299–308. https://doi .org/10.1016/j.enpol.2015.06.008.

Whiteley, Don. 2005. "Gas Exploration Soars in B.C. as Reserves Balloon: [Final Edition]." *Vancouver Sun*, 28 December. Sec. BusinessBC.

Whitton, John, Matthew Cotton, Ioan M. Charnley-Parry, and Kathy Brasier, eds. 2018. *Governing Shale Gas: Development, Citizen Participation and Decision Making in the US, Canada, Australia and Europe*. 1st ed. London: Routledge.

Williams, Laurence J., Abigail Martin, and Andy Stirling. 2022. "'Going through the Dance Steps': Instrumentality, Frustration and Performativity in Processes of Formal Public Participation in Decision-Making on Shale Development in the United Kingdom." *Energy Research and Social Science* 92 (October): 102796. https://doi.org/10.1016/j.erss.2022.102796.

Williams, Laurence J., and Benjamin K. Sovacool. 2019. "The Discursive Politics of 'Fracking': Frames, Storylines, and the Anticipatory Contestation of Shale Gas Development in the United Kingdom." *Global Environmental Change* 58 (September): 101935. https://doi.org/10.1016/j .gloenvcha.2019.101935.

Wilson, Rick K. 2011. "The Contribution of Behavioral Economics to Political Science." *Annual Review of Political Science* 14 (1): 201–23. https://doi.org /10.1146/annurev-polisci-041309-114513.

Wiseman, Hannah J. 2014. "The Capacity of States to Govern Shale Gas Development Risks." *Environmental Science and Technology* 48 (15): 8376–87. https://doi.org/10.1021/es4052582. Medline:24611939

Wolfe, Michelle. 2012. "Putting on the Brakes or Pressing on the Gas? Media Attention and the Speed of Policymaking." *Policy Studies Journal* 40 (1): 109–26. https://doi.org/10.1111/j.1541-0072.2011.00436.x.

Wolsink, Maarten. 2007. "Wind Power Implementation: The Nature of Public Attitudes: Equity and Fairness Instead of 'Backyard Motives.'" *Renewable and Sustainable Energy Reviews* 11 (6): 1188–207. https://doi.org/10.1016/j .rser.2005.10.005.

Wood, James. 2012. "Fracking Causes Water Worries for NDP; Wants Review of Extraction Process." *Calgary Herald*, 22 August.

Wylie, Sara Ann. 2018. *Fractivism: Corporate Bodies and Chemical Bonds*. Durham, NC: Duke University Press.

Yedlin, Deborah. 2015. "Collyer Witnessed Big Changes during Six-Year Term with CAPP." *Calgary Herald*, 13 January. Sec. C.

Younger, Andrew. 2014a. "Andrew Younger Statement on Nova Scotia Hydraulic Fracturing Ban." *Halifax Chronicle Herald*, 3 September. http://thechronicleherald.ca/opinion/1233809-andrew-younger-statement-on-nova-scotia-hydraulic-fracturing-ban.

– 2014b. "From the Desk of Andrew Younger: Hydraulic Fracturing in Nova Scotia." *From the Desk of Andrew Younger* (blog). 4 September. https://web.archive.org/web/20160307070828/http://www.blog.andrewyounger.ca/2014/09/hydraulic-fracturing-in-nova-scotia.html.

– 2014c. "From the Desk of Andrew Younger: Hydraulic Fracking Decision about Social License." *From the Desk of Andrew Younger* (blog). 21 October. http://www.blog.andrewyounger.ca/2014/10/hydraulic-fracking-decision-about.html.

Zaccagna, Remo. 2012. "Ban Urged on Treatment of Waste Water." *Halifax Chronicle Herald*, 21 October. http://thechronicleherald.ca/novascotia/151808-ban-urged-on-treatment-of-waste-water.

Zinn, Jens, ed. 2008. *Social Theories of Risk and Uncertainty: An Introduction*. Oxford: Blackwell Publishing.

Zirogiannis, Nikolaos, Jessica Alcorn, John Rupp, Sanya Carley, and John D. Graham. 2016. "State Regulation of Unconventional Gas Development in the U.S.: An Empirical Evaluation." *Energy Research and Social Science* 11 (Supplement C): 142–54. https://doi.org/10.1016/j.erss.2015.09.009.

Zito, Anthony R., and Adriaan Schout. 2009. "Learning Theory Reconsidered: EU Integration Theories and Learning." *Journal of European Public Policy* 16 (8): 1103–23. https://doi.org/10.1080/13501760903332597.

Zuckerman, Gregory. 2013. *The Frackers: The Outrageous Inside Story of the New Billionaire Wildcatters*. American 1st ed. New York: Portfolio.

Index

Page numbers in *italics* represent tables and figures.

Studies in Comparative Political Economy and Public Policy